WIRELESS BASICS

by HARRY E. YOUNG

Telephony

Div. Intertec Publishing Corp.
55 E. Jackson Blvd.
Chicago, IL 60604

First Edition: January, 1992

Library of Congress Catalog Card Number: 91-068498

ISBN: 0-917845-17

Printed in the United States of America

Foreword

Wireless!

It's one of the most popular terms in telecommunications these days. To some, it connotes an entirely new era and technology. In a sense that is true, but the term has been around for over a century and has remained synonymous with "radio" in many parts of the world. So the new wireless services do represent a progression of technology, and perhaps a new era in telecommunications, but they are still radio-based services that are distant offspring of the early pioneers like Guglielmo Marconi.

There are a variety of wireless services used for telecommunications today and new concepts are emerging. Each provides a vital communications link, but to enable the communication desired by the consumers, these wireless services must be linked with each other and, most of all, with the existing landline telephone network. Without this interconnection, the systems providing these wireless services are analogous to a small group of islands with inter-island air service, but no air service to the "mainland."

This book describes, in very basic terms, various wireless services and their interconnection with the existing landline telephone network, which is known as the *Public Switched Telephone Network (PSTN).* The objective is to give the reader an understanding of the services offered by the wireless carriers and of the interconnection needs of each of those wireless carriers.

The specific wireless carriers that are presented in this book are two-way mobile, cellular, paging, air-to-ground, and the emerging *Cordless Telephone 2nd Generation (CT2)* and *Personal Communications Network (PCN)* concepts. Each description contains a brief history of the service, its radio spectrum requirements, pertinent regulatory considerations, basic system architecture, typical interconnection requirements, numbering needs, and typical call processing procedures.

Two introductory chapters are offered to introduce the reader to some radio and interconnection fundamentals. Each is rather brief because other books exist which are devoted to basic radio and telephony principles. The purpose of these chapters is to explain terms or concepts that are used in later chapters.

The radio chapter includes a review of the radio spectrum. It illustrates the relationship among frequency, antenna size, and range so the reader understands why certain frequencies are more suitable than others for certain applications. It also contains simple explanations of common modulation techniques, like *Amplitude Modulation (AM), Frequency Modulation (FM), Frequency Division Multiple Access (FDMA), Time Division Multiple Access (TDMA),* and *Code Division Multiple Access (CDMA).*

The interconnection chapter begins with a review of the common elements of the PSTN, such as end offices, local tandems, access tandems, and interexchange carriers. An explanation of line-side versus trunk-side connections and the types of signaling used in the PSTN is included. Also presented are the main characteristics of the common interconnection and interface types requested by wireless carriers as well as two services, "Calling Party Pays" and wide area calling plans. Also, since numbering is just as critical to most wireless carriers as it is to the landline carriers, the elements of the *North American Numbering Plan (NANP)* are reviewed.

Finally, there are regulatory considerations for wireless services that must be considered. The *Federal Communications Commission (FCC)* has complete control of the issuance of radio licenses in the United States and exerts a powerful influence on interconnection matters. But the states have jurisdiction on matters that are strictly intrastate in nature. This chapter provides an overview of the major regulatory considerations affecting wireless services.

TABLE OF CONTENTS

The Changing Times

In the last nine years, cellular radio has altered the world of telephony forever. It has changed the basis for the technology. It has changed people's perceptions. And it is likely to change telecommunications as we know it in the future.

For decades, the world has viewed telecommunications as a wired technology, tracing signals through cables, wires and fiber-optics. The infrastructure has been in place to deal with the challenges of that medium for today and for decades to come. However, telecommunications systems in which signals that are based on radio frequency transmission are now taken seriously as a high quality communications medium.

With the presence of wired technology, consumers' mindsets viewed telecommunications as it related to that phone plugged into the wall. However, starting out small, and initially misnamed as a rich man's toy, the cellular phone began to change how people viewed talking to the world at large. It didn't have to be on that phone in the kitchen or hall, plugged into an office wall or at the payphone down the street. It could potentially be anywhere and anytime.

With the changes proffered by the cellular industry, industries and consumers alike are beginning to not only think but believe the concept of reaching people anytime, anywhere, rather than reaching them at a fixed location. Some of these changes are beginning to show up in cellular developments and the first steps of CT-2 and personal communications services.

When this much change occurs in perceptions and shakes the very basis of older, more proven technologies, you know that there is potential. There is power in the instrument. There is indeed a profound desire and need for change.

Rhonda L. Wickham
Editor
Cellular Business

1 RADIO FUNDAMENTALS

INTRODUCTION

All wireless services for voice transmission are based on the use of radio as a transmission medium. But few people concern themselves with the technical aspects of radio because their sole concern is whether the service they are paying for works properly. Besides, radio seems so encumbered by technical jargon, such as frequency spectrum, bandwidth, modulation, and multiplexing. It does seem confusing, but if we can talk to people in spaceships, what could be so technically difficult about talking to your business associate or Uncle Bill who may be only a few miles, or a few blocks, from you?

Radio transmission has many complexities, and obtaining satisfactory performance for conversations that travel only a few miles can sometimes be as difficult as talking to people in space. Major factors affecting the range of radio signals are the *frequency* used to broadcast or receive the message, the length and height of the *antenna,* the *power* used to transmit the signal, and impairments encountered along the transmission path. How many people can utilize a particular radio band is related to the *bandwidth* of the signal and the *multiplexing* technique that is used.

FREQUENCY SPECTRUM

All radio transmissions use electromagnetic waves that result from alternating electric currents flowing through an antenna. A single wave starts with no current flow, increases to maximum flow in the positive direction, reverses the flow until maximum flow at the bottom (or negative) direction is achieved, and finally returns to zero. *Frequency* is defined as the number of these *cycles* that occur in a time period of one second. Figure 1.1 illustrates this concept.

These repetitions, or oscillations, were once known as *cycles per second (cps)* but are now referred to as Hertz per second, in honor of a radio pioneer, Heinrich Hertz. The frequency of these oscillations can range from 1 Hertz (1 Hz) per second to trillions of times per second, and the entire range of frequencies is known as the *frequency spectrum.* Higher frequencies are referred to as kiloHertz (which is one thousand Hertz, 1 kHz), megaHertz (one million Hertz 1, MHz), gigaHertz (one billion Hertz, 1 GHz), or teraHertz (one trillion Hertz, 1 THz).

Typical Radio Wave
FIGURE 1.1

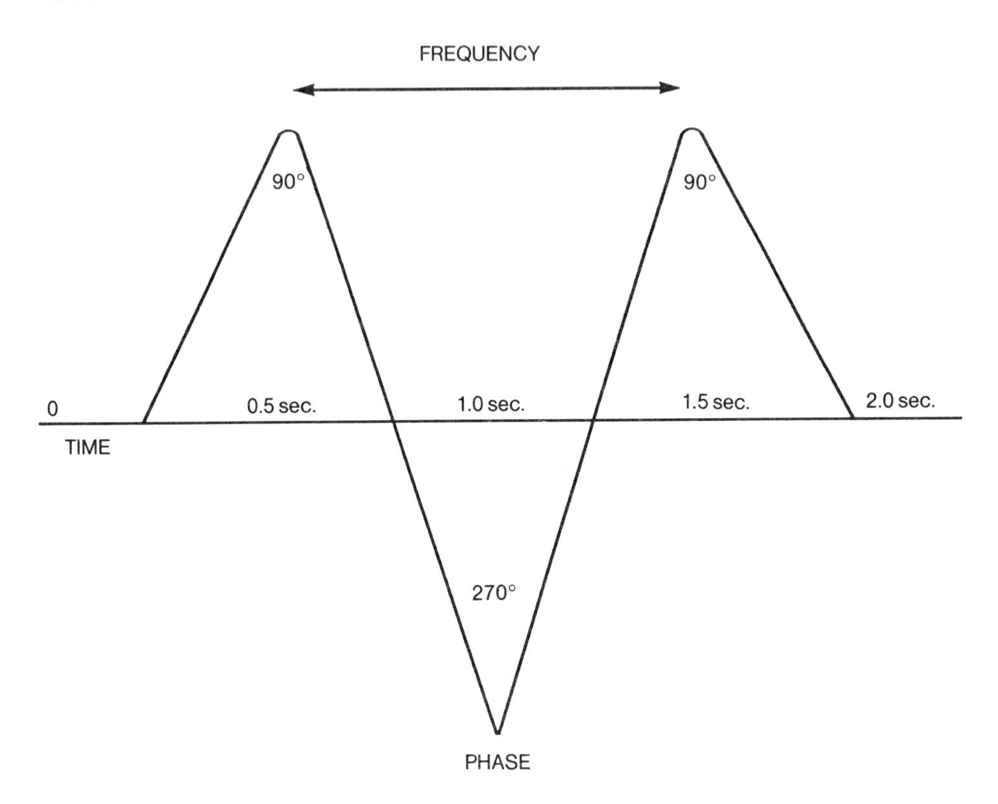

Frequency Bands and Bandwidth

Only a portion of the entire frequency spectrum is suitable for radio transmission. It is generally considered to extend from about 30 kHz to 300 GHz. Some radio transmission is possible with frequencies as low as 72 Hz, but in reality, 100 GHz is the practical upper limit with current technology. The radio spectrum is divided into categories called *frequency bands* which have descriptive terms such as VHF (very high frequency) and UHF (ultra high frequency). These bands and their frequency ranges are shown in Table 1.1.

The size of each frequency band is determined by its *bandwidth,* which is merely the difference between the highest and lowest frequencies in a particular band. For instance, the VHF band has a bandwidth of 270 MHz because it lies between 30 and 300 MHz. As we shall see, bandwidth is also used with respect to individual radio channels that are utilized within a frequency band.

FREQUENCY AND WAVELENGTH

A relationship exists between the frequency of a radio signal and the length of its radio wave, or wavelength. Since radio waves travel through the air at approximately the speed of light, the distance that is traveled during one complete cycle is shorter for higher frequencies than it is for lower frequencies. Therefore, the higher frequencies involve smaller wavelengths than lower frequencies. Wavelengths have tremendous variations in size. For example, a 30 kHz signal that is used to contact submarines underwater has a wavelength of about 6.2 miles while the wavelength of an 800 MHz cellular signal has a wavelength of about 13 inches. Since the speed of light is approximately 186,000 miles per second, the wavelength can easily be calculated by dividing 186,000 by the frequency of the signal. Wavelengths are more commonly expressed in meters, which can be obtained by dividing 300,000,000 by the frequency (in Hertz) of the signal.

Table 1.1

Frequency Band	Frequency Range	Examples Of Uses
Extremely Low Frequency (ELF)	Less than 3 kHz	Naval communications systems
Very Low Frequency (VLF)	3-30 kHz	Omega navigation system
Low Frequency (LF)	30 kHz-300 kHz	LORAN-C navigation system
Medium Frequency (MF)	300 kHz-3 MHz	AM broadcasting
High Frequency (HF)	3 MHz-30 MHz	FM broadcasting
Very High Frequency (VHF)	30 MHz-300 MHz	FM broadcasting, 2-way mobile
Ultra High Frequency (UHF)	300 MHz-3 GHz	TV broadcasting, cellular, paging
Super High Frequency (SF)	3 GHz-30 GHz	Microwave transmission
Extremely High Frequency (EHF)	30 GHz-300 GHz	Satellite systems

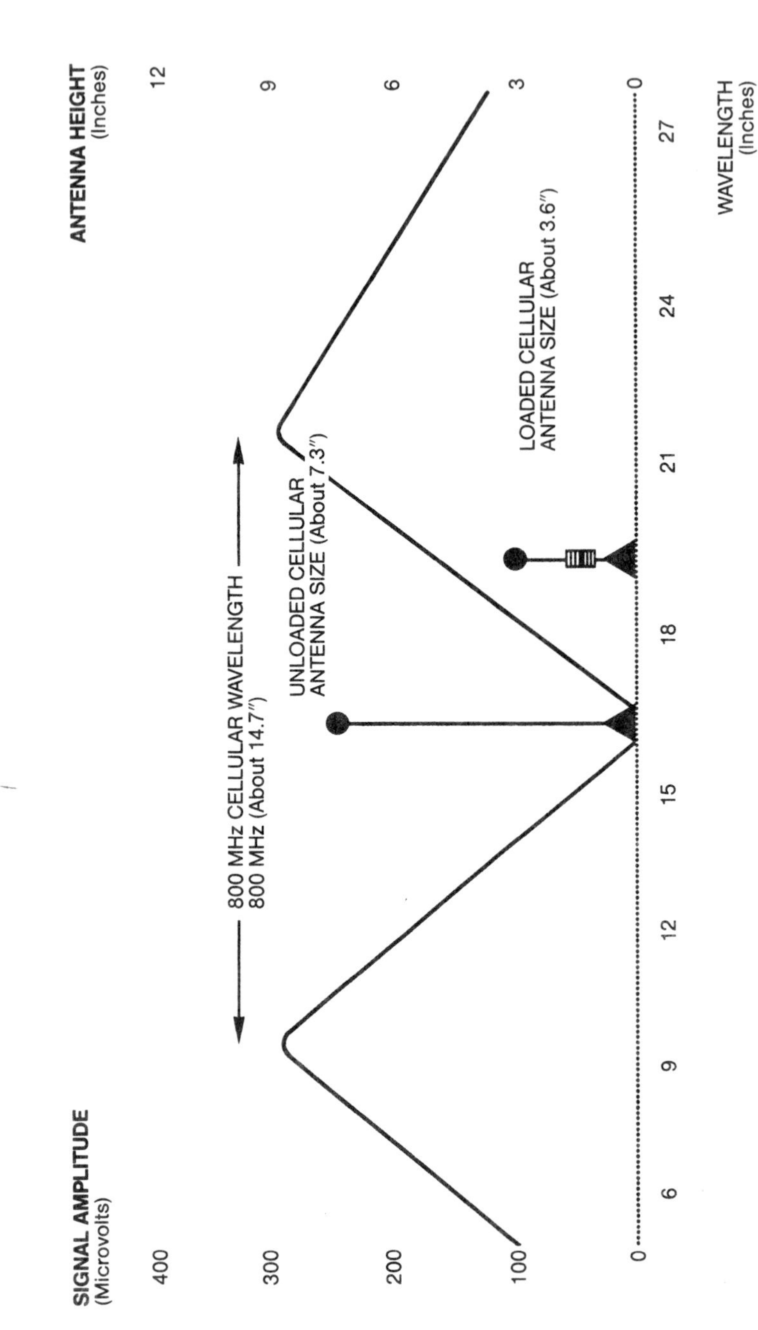

Wavelength Versus Antenna Size
FIGURE 1.2

WAVELENGTH AND ANTENNA SIZE

Antennas serve to radiate or receive a radio signal. In order to maximize the power that is radiated, or conversely, received, antennas have optimal sizes that are related to the wavelength of the signal. Basically, the higher frequencies, with their shorter wavelengths, require smaller antennas. An optimum antenna is usually one that is one-half the size of the wavelength. However, antennas mounted vertically on metal surfaces or on the earth, generally have two conductive surfaces. One, known as the "real" section, is a physical conductor. The other, called the "image" conductor, exists only in an electrical sense. For this reason, an antenna that provides excellent characteristics for a particular frequency has a size that is typically only one-fourth the size of the wavelength and is known as a "quarter-wave" antenna. The physical size of an antenna can be further reduced by the addition of an electrical device called an *inductor.* This technique, called loading, can reduce the size of a typical citizens band (CB) from one that would normally require a 9-foot length down to a more practical 3-foot size. Figure 1.2 shows the relationship between signal frequency and antenna size.

Antenna Gain

Antennas often have a "gain" rating expressed in *decibels.* This measurement is a logarithmic measure of the ratio between two signal powers, typically associated with the antenna in question, versus a reference antenna. This "gain," or increase in signal strength, is not due to any amplification of the signal. Instead, the gain represents the difference in signal strength between what would be received from an isotropic antenna, which theoretically radiates a signal uniformly in all directions, versus a directional antenna that concentrates the signal in a specific direction. Likewise, an antenna that provides gain when transmitting provides the same gain when receiving. It is common to rate radio stations that use gain antennas in terms of *effective radiated power (ERP).* A transmitter that produces one watt of power, sending through an antenna which has a gain of ten, gives an ERP of 10 watts.

PROPAGATION CHARACTERISTICS OF RADIO WAVES

Because the sizes of wavelengths vary, radio signals propagate differently through space. Some are well suited for transmissions involving very long distances while others are limited to short ranges. Typically, the higher the frequency, the shorter the distance the signal will travel. Frequencies with long wavelengths tend to follow the curvature of the earth, which is why the *very low frequency (VLF)* and *low frequency (LF)* bands are especially suitable for global transmission and long-range navigation purposes. Signals below 30 MHz, primarily those in the *medium frequency (MF)* band, travel upward into the sky, but some of their energy is reflected back causing the wave to "skip." This "skipping" effect permits long-range, but not always reliable, communication. In the *very high frequency (VHF)* range and above, radio signals do not bend, and those that are transmitted into space are not reflected back. These frequencies are useful for line-of-sight communications, but are not useful for long distance communications unless they are aimed at a satellite which then retransmits them back to earth.

Radio propagation is also affected by the power levels used to transmit the signal. The transmitted power level of radio signals is expressed in *watts,* as with household appliances, except that the wattage value is usually much lower. A relationship also exists between power levels and frequency in that higher frequencies, with their smaller wavelengths, require higher power levels to transmit a signal the same distance as lower frequencies. The maximum transmit power for some television stations in the very high frequency (VHF) band, for example, is 100 watts, while TV stations operating in the *ultra high frequency (UHF)* band have a maximum transmit power of 5000 watts.

The signal strength of a radio wave diminishes rapidly as it moves away from the transmitter antenna. With absolutely no obstructions, this reduction would be the inverse of the square of the distance. For example, a received signal that is 10 milliwatts (1/100th of a watt) one mile from the transmitter will be only 2.5 milliwatts two miles from the transmitter and only 0.1 milliwatts (or 1/10000th of a watt) at ten miles. In reality, due to obstructions (buildings, etc.) or absorptive material (trees), the loss can be far more severe, approaching the inverse of the 5th or 6th power of the distance.

Radio signals do not arrive at the receiver with constant intensity. Instead they are subject to *fading.* AM broadcast signals, discussed below, vary in level because the signal traveling along the earth, directly from the transmitter, may clash with the same signal reflected from the upper layers of the earth's atmosphere. The *phase* of these two signals, which pertains to the position of the sine wave in Figure 1.1 at an instance of time, may shift with respect to one another. This phase shift can alternately cancel or reinforce the signal. Fading can also occur with line-of-sight transmission, like FM radio or cellular, because of interference between the signal received directly from the transmitter and a signal that is reflected from buildings or land features. Fading occurs with hand-held radios as the user moves between zones where the signal adds or subtracts.

Directly related to the effects that cause fading is the effect of multipath transmission: the arrival of a signal over both a direct and a reflected path, or over two reflected paths. In addition to changing the level of a received signal, this effect distorts its quality.

For digital radio systems, like those being tested for cellular applications, multipath transmission results in a given information pulse being received several times, slightly spread apart in time because of differing distances traveled by the direct and reflected waves. The pulse does not have the same intensity as received over the several paths. The undesirable effects of this *dispersion* or *delay spread* are controlled by suitable system design.

Radio signals are subject to interference from several sources of noise. Interference from natural causes, mainly lightning, is strong below 1 MHz, but tapers off and becomes minor above 30 MHz. Noise from power lines can be severe in the same frequency range. Automobile ignition systems produce sizeable interference, but that noise falls off above about 300 MHz. The result is that, for many wireless applications discussed in this book, most of the noise experienced is actually generated in the user's receiver itself.

MODULATION SCHEMES

Radio waves transport information from one location to another. In a sense, they are analogous to railroad cars in that they are valuable only when they are carrying something. The frequency of the radio wave providing the transport is known as the *carrier frequency.* The information to be carried is mixed with this carrier frequency by a process known as *modulation.*

Modulation is necessary because the intelligence of the signal, voice for instance, is usually of such low frequency that it cannot be readily radiated into space. Two widely used forms of modulation are *amplitude modulation (AM)* and *frequency modulation (FM).* Others are *quadrature amplitude modulation (QAM), phase modulation (PM),* and *pulse code modulation (PCM).*

It is common to use combinations of modulation techniques. For example, FM stereo broadcasting uses a combination of AM and FM. Digital radio systems convert voice signals to pulse code modulation, then use QAM or PM to convey the pulse stream on the radio signal.

Amplitude Modulation (AM)

Amplitude modulation was the first form of modulation and is still very common, particularly for commercial radio stations. If one tunes a radio to 1000 kHz (or 1 MHz), the carrier frequency is 1000 kHz. Voices that are transmitted over this frequency are mixed with the 1000 kHz carrier frequency. The amplitude, or size of the sine wave of the carrier frequency, increases or decreases at the same rate as the voice signal.

Because AM varies the amplitude of the carrier wave, it is quite susceptible to unwanted signals (noise) being included with the intelligence as the signal is radiated. When the carrier wave is removed at the radio receiver (demodulated), the noise remains with the intelligence, a very undesirable characteristic. Due to this limitation, particularly from the ignition systems in vehicles, AM was not well suited for early mobile applications. It is still used in some applications, such as aircraft, where the noise sources are controlled separately.

Quadrature Amplitude Modulation (QAM)

A special variant of amplitude modulation, called quadrature amplitude modulation (QAM), is used for digital systems like microwave transmission systems and AM analog stereo broadcasting. QAM generates two carrier waves of the same frequency, but separated by a quarter-wave in time. These waves can be individually modulated with a signal and separated at the receiver, thus doubling the capacity of the radio channel.

Frequency Modulation (FM)

In order to overcome the inherent noise problems with AM, frequency modulation was developed. With FM, the frequency of the carrier wave, not its amplitude, is varied by the intelligence of the signal being carried (voice, music, etc.). Because the frequency is varied, it is largely unaffected by any noise that is encountered after the signal is radiated. FM is used where noise mitigation (mobile applications) and fidelity of the signal (stereo FM broadcasting) are important.

Phase Modulation (PM)

Phase modulation is related to frequency modulation; with it, the carrier wave is advanced or delayed by a fraction of a cycle to convey either analog or digital signals.

Pulse Code Modulation (PCM)

Digital radio systems convert analog voice signals into a series of on-off pulses. They sample the voice wave at periodic instants, typically 8000 times per second, then measure the height of the wave and send it by a binary code that indicates that height. At the receiving end, the binary code is converted back to an analog voice waveform.

MULTIPLEXING TECHNIQUES

In order to increase the capacity of the radio band that is used for a particular application, like cellular service, *multiplexing* techniques are used. Multiplexing can be compared to increasing the traffic capacity of a bridge over a river. The existing traffic lanes of the bridge can be narrowed to accommodate more lanes, or perhaps an entire new level can be added to the bridge. In any case, the bridge is no wider than before, but it can now carry more cars over its span. Many forms of multiplexing exist but three popular forms are *frequency division multiple access (FDMA), time division multiple access (TDMA),* and *code division multiple access (CDMA).*

Associated with multiplexing is the *bandwidth* each channel, or circuit, occupies within a given band. Bandwidth was previously discussed when frequency bands were defined. It is simply the difference between the highest and lowest frequencies in the band. The same concept applies to channels on a smaller scale. Channel bandwidth is like the width of the traffic lanes in the bridge; they can be narrow or wide.

In each of these multiplexing schemes, the term *multiple access* is used. This means the radio channels are shared by the users instead of assigning each user an unique frequency. In this context, multiple access is synonymous with the term *trunking*.

Frequency Division Multiple Access (FDMA)

Existing cellular systems use FDMA multiplexing, which divides the total band allocated to a cellular operator (25 MHz) into discrete channels. Since each channel has a bandwidth of 30 kHz, the system has a total of 833 channels available. Each conversation requires the use of two pairs of frequencies, so there are 416 pairs of frequencies available to each operator, each of which can potentially be assigned to a cellular user at any given time. Mobile equipment that utilizes FDMA is less complex than that used with other multiplexing techniques and is generally less expensive. However, because each channel requires the use of a separate transmitter and receiver, FDMA needs considerably more equipment at the non-mobile, or *base station* site. FDMA can be used with analog or digital transmission systems.

Time Division Multiple Access (TDMA)

With TDMA, each radio channel is divided into time slots. Each individual conversation is converted to a digital signal which is then assigned to one of these time slots. The number of time slots per radio channel can vary, because it is a function of the system design. There are at least two time slots per channel, and usually more, which means that TDMA has the potential of serving several times more customers with the same amount of bandwidth as FDMA techniques.

TDMA is a more complex system than FDMA because the voice has to be digitized, or coded, then stored in a buffer for assignment to a vacant time slot, and finally transmitted. Consequently, signal transmission is not continuous and the transmission rate must be several times greater than the coding rate. Also, because more information is contained within the same bandwidth, TDMA equipment must have more sophisticated techniques of equalizing the received signal in order to maintain signal quality.

Code Division Multiple Access (CDMA)

CDMA was developed by the military to prevent communications channels from being detected, and then jammed, by the enemy. The signal intelligence, e.g., speech, is converted to a digital signal which is then mixed with a random-like code. The total signal, speech plus the random-like code, is then transmitted over a wide band of frequencies through a technique called *spread spectrum*. It may be helpful to visualize this concept as looking from an overpass to a stream of cars passing below on the freeway. The cars appear to be in random order but the intelligence could be represented by particular types of cars (Fords, Nissans, etc.) while the random-like code could be represented by a combination of cars painted with particular colors that is randomly changed. For example, a Ford surrounded by three blue cars could signify a digital "1" while a Nissan surrounded by a group of red cars could indicate a digital "0." Since these sequences could be rapidly changed, it would be difficult for an observer on the overpass to detect any intelligence from this seemingly random stream of cars. Likewise, it is difficult for an outside party to detect the radio signal and intelligence when it is spread over a wide bandwidth in a random-like manner. There is the further potential for multiple radio systems to share the same spectrum with only limited interference.

Unlike FDMA and TDMA, the spread spectrum transmission used by CDMA requires channels that have a relatively immense bandwidth (perhaps as much as 10 MHz). However, it is theorized that CDMA could accommodate perhaps 20 times as many users for the same amount of total bandwidth as FDMA can. Until now, CDMA has been used for specialized purposes with small numbers of users. A major question to be answered is whether one service, with tens of thousands of users employing CDMA devices, can function and not interfere with other wireless services.

FREQUENCY INTERFERENCE AND COORDINATION

Interference with radio signals can be caused by other man-made radio signals. Two major sources of this interference are *adjacent channel interference* and *co-channel interference.* In both cases there are measures that can be used to reduce this interference. These measures can include equipment performance standards, proper channel spacing, physical separation of users, and coordination of frequency use within a given area.

Adjacent Channel Interference

If radio channels are too closely spaced, or if the electronic filters are inadequate, signals from one radio channel can interfere with an adjacent channel. These adjacent channels operate on different frequencies but they are close enough so that interference can occur. This problem is alleviated by proper system design and manufacturing equipment to specific tolerances, as well as requiring geographic spacing between users on the adjacent channels.

Co-Channel Interference

Unlike adjacent channel interference, *co-channel interference* occurs when signals from one source interferes with another when channels utilize the same frequencies. This phenomenon is reduced to acceptable levels by physically separating radio transmitters from one another, and by limiting the power levels of the transmitter. Stations having antennas at unusually great heights often have their power limited to reduce their potential for causing interference. In a cellular system, the reuse of frequencies is predicated on engineering the system that physically separates cells using the same frequencies in a predictable manner. Other radio systems must be separated by specific mileages that are specified by the Federal Communications Commission (FCC). The FCC requires radio operators to submit data indicating potential interference. This task was formerly done with laborious manual calculations. Today, there are computer programs available to determine if existing systems will be affected by the entry of a new radio signal.

2 INTER-CONNECTION: TECHNICAL FUNDAMENTALS

Wireless services are growing at an explosive rate, but the vast majority of telecommunications users are still customers of the traditional, land-line telephone company. However, to make their services attractive to the majority of customers, wireless carriers need to connect their networks to the telephone companies' network, which is also referred to as the *Public Switched Telephone Network* or *PSTN.* So, an understanding of the PSTN and the types of interconnection arrangements available to wireless carriers is essential.

PSTN COMPONENTS

Technology and the Bell System divestiture of 1984 have radically changed the PSTN. Beginning in the 1950s, the PSTN was a five-level hierarchical network that was based wholly on *analog* facilities and analog switching machines. With the introduction of digital facilities in the 1960s and digital switches in the 1970s, the network had evolved to a point in 1984 where all long-distance switches and some local switches were digital. At that time, the growth of analog facilities had essentially stopped and wide replacement of existing facilities with digital technology was under way. The changing economics of the network led to considerable "flattening" of the long-distance hierarchy. However, at divestiture the five-level concept remained the theoretical model. The lowest switching machine in this pre-divestiture hierarchy was the *end office.* This office, which was also known as a Class 5 office, provided connections between subscribers in a defined local calling area. These Class 5 offices also had connections to *toll centers,* known as Class 4 offices, which connected to other end offices in distant locations. The *toll centers* handled *toll* traffic only. They in turn had connections to other toll offices that were either at the same level (Class 4) or higher in the hierarchy. The higher offices were labeled *primary centers* (Class 3), *sectional centers* (Class 2), and *regional centers* (Class 1). Strict transmission parameters were established for the circuits linking each office. These parameters, such as loss, noise, and balance, were progressively more rigid moving from Class 4 to Class 1 offices. Also, there were limits on how many links could be utilized in a particular call and on the routing of the calls between these offices.

Traditional Components of the Public Switched Telephone Network
FIGURE 2.1

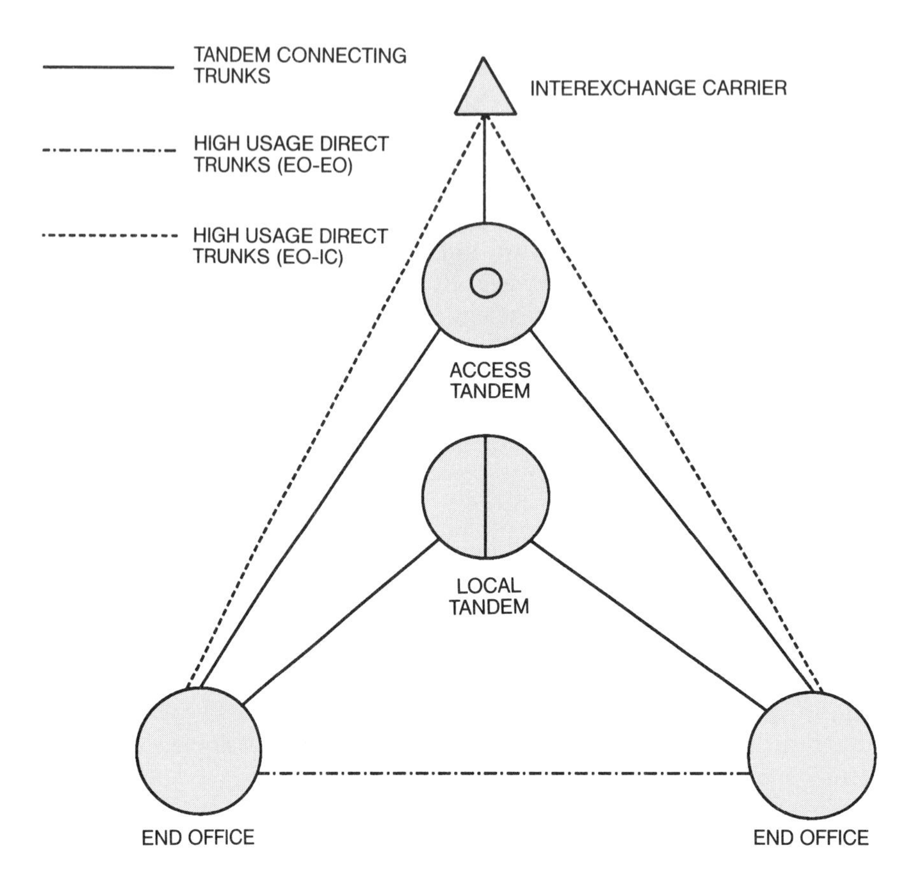

With the introduction of digital switches and digital facilities, some of the transmission considerations that made a hierarchical network imperative were no longer factors. This, coupled with the divestiture, has led to the use of non-hierarchical networks by many *interexchange carriers (ICs)* and a two-level hierarchy by the *local exchange carriers (LECs).* The two-level network is composed of *access tandems* and *end offices.* In addition, some of the LECs use *local tandems* as well, but these are not considered a separate hierarchical layer. A typical two-level hierarchical network used by the LEC is shown in Figure 2.1.

End Offices

End offices provide access to the PSTN for customers in a given geographical area. In basic terms, the customer normally gets dial tone from an end office and is able to make calls to other customers. The size of an end office can vary from a few dozen to tens of thousands of subscribers. Each end office is linked to other end offices within the same *local calling area* either by connections to a local tandem or through a combination of direct trunks to other end offices and local tandem connections. Toll calls are routed to the access tandem for transport by an IC or by the LEC, depending upon the call destination.

Local Tandems

Local tandems do not provide service directly to subscribers because their purpose is to link other switching machines, i.e., end offices, and not to link customer equipment such as telephones. Local tandems have connections to other end offices and are normally used when direct connections between end offices are not economically justified, or they handle overflow if all direct connections between two end offices are in use. Traffic in local tandems is confined to a defined local calling area.

Access Tandems

At divestiture, service areas known as *local access and transport areas (LATAs)* were established. Because the LEC's business is confined to *intraLATA* operation, access tandems were created at divestiture to provide an entry point into a LATA for the ICs, whose business is primarily *interLATA*. The transmission requirements are more rigid for an access tandem than an end office so it represents the higher ranked office in this new two-level hierarchy. However, in relationship to an IC's switch, it is considered to be the lower ranked office.

LINE-SIDE VERSUS TRUNK-SIDE CONNECTIONS

When discussing connections to a central office, the terms *line-side* and *trunk-side* are often used, but the meaning of these terms is not always understood. In simple terms, line-side central office connections connect customer locations with the switch while trunk-side connections connect switching entities, usually central offices, with each other. This general definition is sufficient for most uses, but as with most generalizations, there are exceptions.

Line-side connections are sometimes referred to as *loops* because the physical plant that is used to connect the customer location with a central office is called a loop. To further confuse matters, *trunks* are always "trunks" even though they may occupy a portion of the same physical plant used to provide loops.

Figure 2.2 illustrates the line-side and trunk-side connection concept as it applies to the PSTN.

Another source of confusion is the fact that some line-side connections are referred to as trunks even though they are truly line-side connections. A prime example of this mis-labeling is PBX trunks, which are line-side connections used to connect a PBX with a central office. To

Line-Side Versus Trunk-Side Connections
FIGURE 2.2

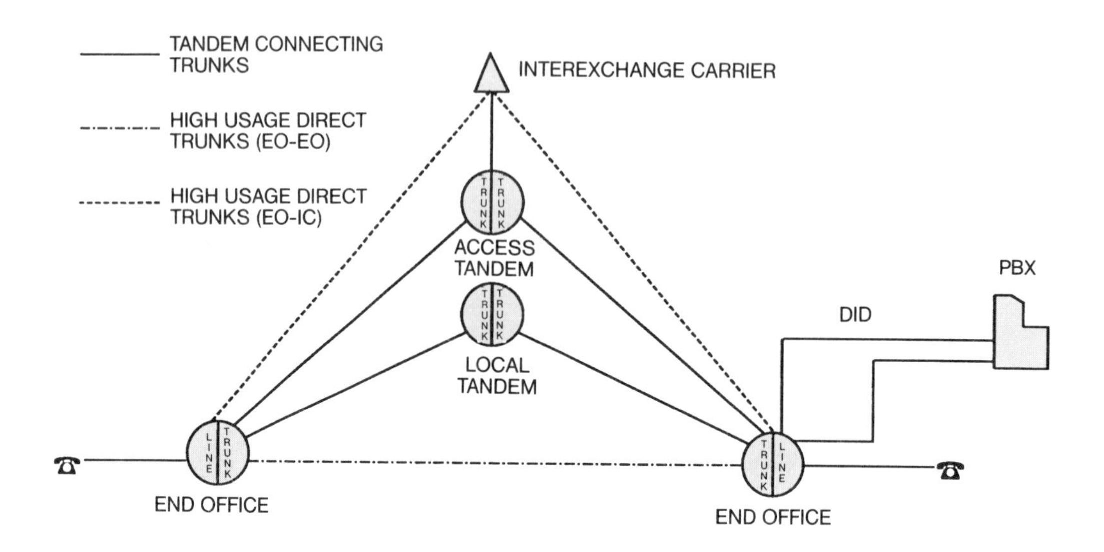

further obfuscate the distinction, *Direct Inward Dialing (DID)* circuits use trunk-side protocols and equipment, but are often considered line connections.

There are significant differences between line-side and trunk-side connections with respect to the signaling techniques that are employed and the transmission requirements for each. For example, telephone sets will not work when directly attached to a trunk-side connection because telephone sets are not designed for the signaling protocols used for trunk-side connections. Furthermore, the transmission requirements for trunks are far more stringent than those for lines or loops.

BASIC SIGNALING ELEMENTS

Signaling in the telephone network involves a number of aspects and there are several methods of indicating the same function, depending upon the type of equipment used and whether the connection is a line-side or a trunk-side connection.

The signaling elements can be associated with the circuit, meaning they are carried on the same facility as the voice, or a separate, *common channel signaling (CCS)* path may be used. If the signaling is circuit-associated, it may use inband or out-of-band signaling. Inband signaling uses the voice path itself while out-of-band uses separate voice and signaling paths. Signaling from a normal pushbutton phone is an example of inband signaling, while the CCS networks used for signaling between some switching offices uses out-of-band signaling.

There are six basic signaling functions that are used within the network. These functions, which are summarized in Table 2.3 are address, supervisory, alerting, call progress, control, and test signals.

Address Signals

These signals convey call destination information, or the digits dialed by the calling party. Since these digits are actually electrical pulses, the term *address pulsing* is used to indicate the type of signaling used for this function. Common types of address pulsing are *dial pulse (DP),* which is the oldest and slowest form of address pulsing, *dual tone multifrequency (DTMF),* which are the signals from a pushbutton phone, and *multifrequency (MF),* which is used exclusively for trunk signaling.

Supervisory Signals

This function merely indicates the status of the line or trunk. It is associated with the terms "off-hook" and "on-hook," which indicate a busy or idle condition. Common types of supervisory signals are *loop, reverse battery,* and *E&M supervision.*

From the use of the term loop, one may deduce that this type of supervision is used for line-side connections. Loop supervision can be

Table 2.3

Five Basic Signaling Elements

ELEMENT	TYPES
ADDRESS	DIAL PULSE (DP) DUAL TONE MULTIFREQUENCY (DTMF) MULTIFREQUENCY (MF)
SUPERVISORY	LOOP (Loop Start or Ground Start) LOOP (Reverse Battery) E&M
ALERTING	20 Hz Ringing
CALL PROGRESS	AUDIBLE RINGING DIAL TONE BUSY SIGNALS RECORDED ANNOUNCEMENTS
CONTROL	TOLL DIVERSION
TEST	LOOP RESISTANCE TRANSMISSION TESTS

Basic Loop Supervision
FIGURE 2.4

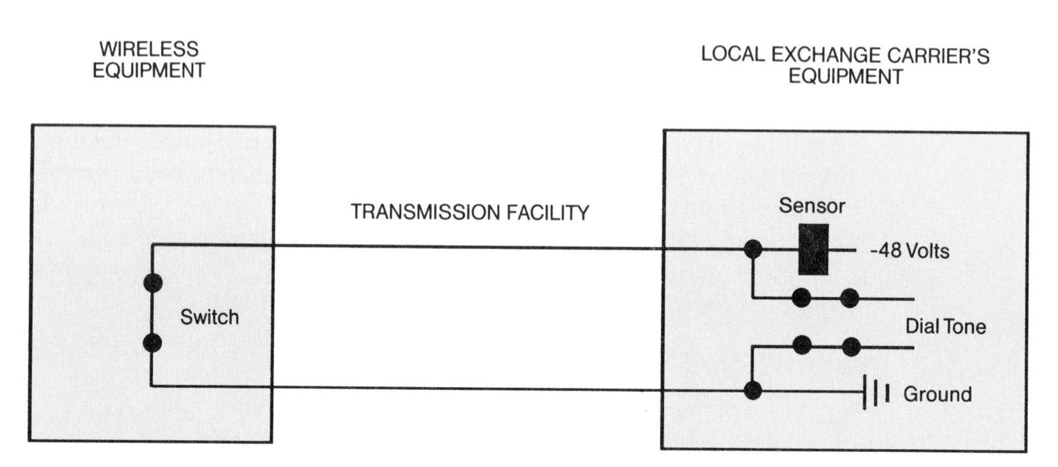

A closure (off-hook) provided by the wireless carrier's equipment completes the loop for the circuit in the local exchange carrier's equipment and provides dial tone.

accomplished by providing a loop closure, as shown in Figure 2.4, which is known as *loop start.* Other forms of loop supervision include applying a ground to one side of the line *(ground start)* or reversing the line polarity, which is called *reverse battery.* Reverse battery is used for trunk-side connections but is restricted to one-way trunks.

E&M supervision, as depicted in Figure 2.5, may be used for one-way or two-way trunk-side applications. "E&M" is not an acronym but a term presumably derived from lead designations on schematic drawings for telephone equipment. Using E&M signaling, trunk seizures are controlled by placing a voltage, or *battery* on the M lead, which causes the E lead at the other central office to change its state from an electrical open to an electrical ground condition. The E&M leads are not separate transmission paths between the offices but instead are conveyed either by signaling bits in a digital signal or tones if an inband signaling method is used.

Another function of supervisory signals is to control the outpulsing of the address signals when trunk-side connections are used. The most common form of address outpulsing control is *wink start,* which is a very short on-hook/off-hook signal that indicates the receiving switch is ready to receive the address digits.

Alerting Signals

These signals alert the called customer, equipment, or operator of call receipt. Alerting signals often use 20 Hz ringing to alert a customer connected with a line-side arrangement.

Call Progress Signals

In order to keep the person making a call informed of the various events that occur during the course of a call, call progress tones or signals are used. For example, dial tone is used on line-side connections to indicate that a line is available for use. Audible ringing is provided to the calling party to indicate ringing has begun. A busy signal occurs when the called party's line is in use or if all trunks to the called party's end office are busy. If a call is made to a number that is not assigned, the call is routed to a vacant number announcement.

Control Signals

Control signals are used for special auxiliary functions that are beyond the *Point Of Termination (POT).* Toll diversion, which restricts calls from some equipment to a local calling area, is an example of a control signal.

Basic E & M Supervision
FIGURE 2.5

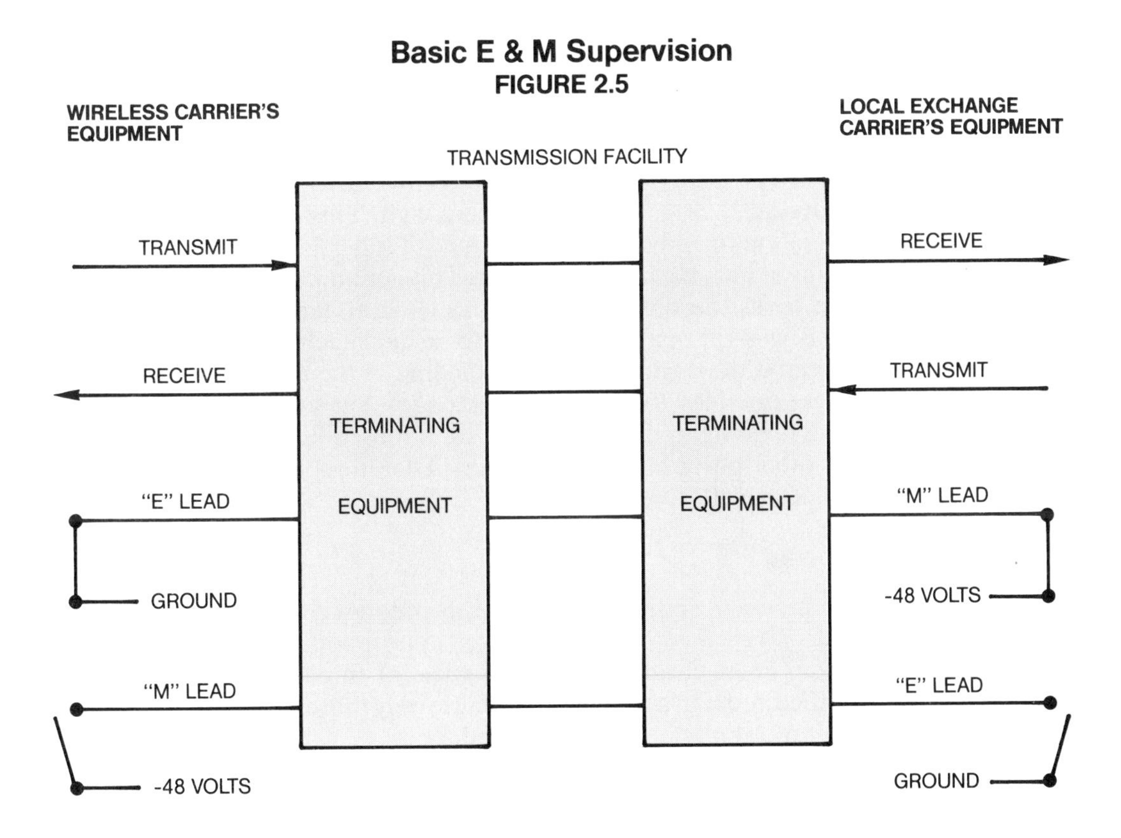

In this example, placing -48 volts on the "M" lead at the LEC location causes a ground on the "E" lead at the wireless carrier's location signifying a trunk seizure. Note that the "M" lead at one location is connected to the "E" lead at the other location.

Test Signals

These signals verify circuit conditions that are not directly associated with the customer's actions. Many switching machines, for example, conduct a low line resistance check that identifies loop irregularities that could cause billing problems. Automated transmission tests are made on trunk-side connections to ensure the trunks meet the transmission objectives.

NORTH AMERICAN NUMBERING PLAN (NANP) ELEMENTS

During the 1940s, the possibility was foreseen of telephone users dialing on a nationwide basis. To enable calls to be routed nationwide, which was called *Direct Distance Dialing (DDD),* a numbering plan was developed for most of North America that assigned a 10-digit address to a particular station. This plan, known as the *North American Numbering Plan (NANP),* is primarily a geographic based plan that permits destination code routing based on the 10-digit address of the called station. The 10-digit address is composed of a 3-digit *Numbering Plan Area (NPA)* code, a 3-digit *central office code (NXX),* and a 4-digit *station code (XXXX).*

Numbering Plan Area (NPA) Codes

Currently, the digits assigned for these codes are restricted to 3-digit combinations that have a "1" or a "0" as the second digit. This allows most switching machines to quickly determine if the call is destined for another NPA by looking at only the first three digits. This action, called a 3-digit translation, reduces the time needed by a switching machine to process a call. This was especially important for older mechanical switches. It also increased the customer's understanding of the dialing plan. This arrangement provides 160 possible NPA codes, but some of them, specifically those ending in the digits "11," are not used because they are reserved for other purposes (911, 411, etc.). Therefore, a total of 152 usable NPA codes results from this plan.

NNX And NXX Codes

Initially, the central office and station portion of the code used a two-letter, five number (2L-5N) arrangement that resulted in telephone numbers like PA6-5000, or "Pennsylvania 6-5000." The use of this format allowed 540 central office codes in a given NPA because only those codes that could result in a central office name were used.

In the 1960s, the demand for telephone numbers was so great that a new arrangement using only numbers was introduced. In this scheme, the codes use a NNX-XXXX format whereby "N" is any number from 2 through 9 while "X" is any number from 0 through 9. This prevented central office codes from having a "1" or "0" as the second digit so central office codes would not conflict with the NPA codes. This made an additional 100 central office codes available since it was no longer necessary to have the digits "spell" a name.

By the 1970s, number demand again had increased so that the format was changed to allow the use of a "1" or a "0" as the second digit in a central office code. This format, called "NXX," further expanded the availability of central office codes to a total of 792 because it allowed for use as central office codes the 152 codes normally reserved for NPA codes. When these NPA codes are interchangeable with central office codes, the switching machines must either do a 6-digit translation, or wait for a finite period of time for additional digits, to determine if the call is addressed to another NPA.

Interchangeable NPA Codes

Because the demand for telephone numbers has continued to increase, the supply of NPA codes in their present NXX format is almost exhausted. Consequently, the use of interchangeable NPA codes is scheduled for about 1995. This will result in 792 codes available for NPA use instead of the existing 152, but requires each switching machine that uses the PSTN to be able to process calls using this interchangeable NPA format.

Service Access Codes (SAC)

These codes, which use the N00 format, are non-geographic. Only three are presently assigned, which are the 700, 800, and 900 series. The 700 series is for use by interexchange carriers only. Numbers in the 800 series are traditionally used for services where the called party, rather than the calling party, pays for the call. The 900 series is used for informational or entertainment services where the calling party usually pays an additional rate to the entity providing the 900 service.

N11 Codes

Codes in this N11 format are used for different purposes. Some, like 411 for Directory Assistance or 911 for Emergency, have become fairly standard. Other code combinations are used for local purposes only.

North American Numbering Plan Administration

Since the divestiture of the Bell System in 1984, Bell Communications Research (Bellcore) has been responsible for administering the North American Numbering Plan. As part of this responsibility, Bellcore assigns NPA and service access codes. Assignment of NXX codes within a given NPA is the responsibility of the Numbering Administrator for that particular NPA, typically a person from a major local exchange carrier. N11 codes are also locally assigned by the local exchange carrier.

INTERCONNECTION TYPES AND INTERFACES USED BY WIRELESS CARRIERS

Basically, there are six types of interconnection arrangements available from the local exchange carriers for use by the wireless carriers. The type of connection used depends upon the wireless system, the intended application, and, in some cases, regulatory considerations. Five of these connection types provide access to the local exchange carrier's switching machines, while the last is a private-line service used to link locations within the wireless carrier's own network. Figure 2.6 illustrates the various interconnection types.

Dial Line Connections

These are the most basic and oldest type of interconnection used by wireless carriers. Dial lines are two-wire line-side connections from an end office that are just like the connections used for business and residence lines. They are two-wire circuits that may be used on a one-way

Alternatives for Interconnection to the Public Switched Telephone Network
FIGURE 2.6

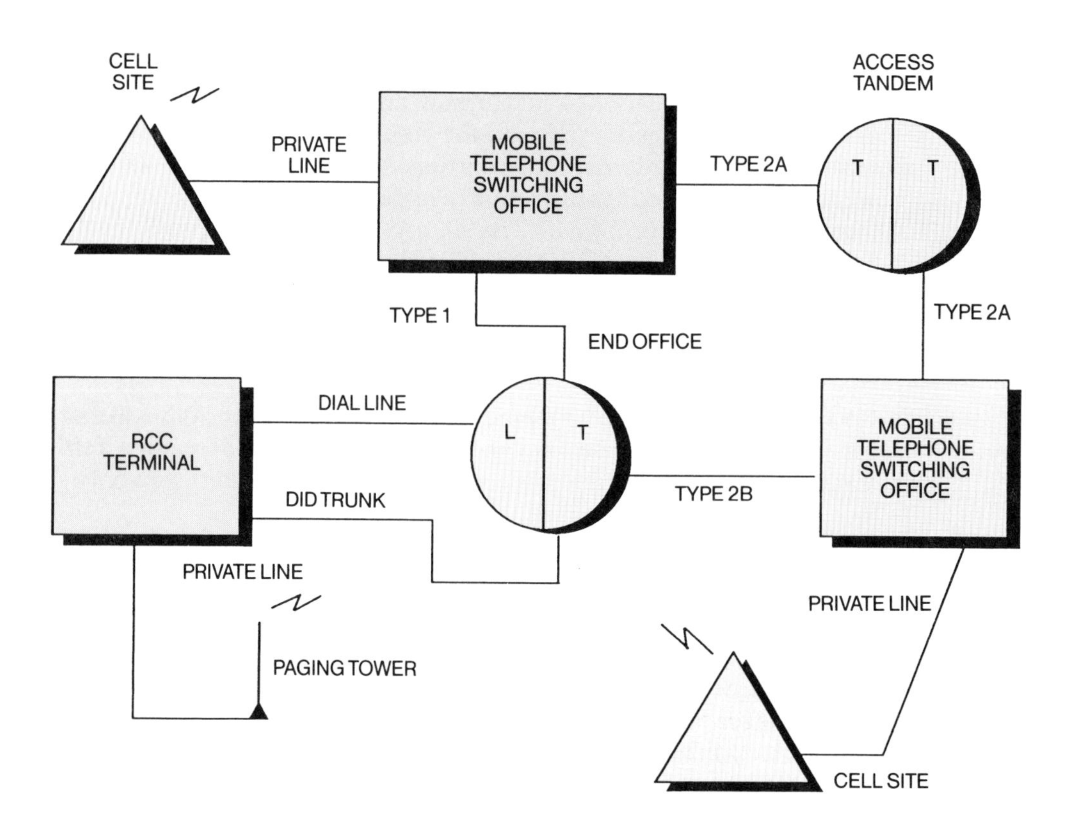

or two-way basis and are available from every type of switching machine used by the local exchange carrier as an end office.

In the outgoing direction (WC-LEC), dial lines provide dial tone upon a seizure that uses loop supervision. This loop supervision can be controlled by loop start or ground start, meaning the wireless carrier provides a loop closure or electrically grounds one side of the loop to indicate a seizure. After line seizure, the digits are forwarded to the end office using either dial pulse (DP) or dual tone multifrequency (DTMF) address pulsing. For incoming calls (LEC-WC), a 20 Hz ringing signal is sent from the end office as an alerting signal.

Like other line-side connections in an *equal access* equipped end office, the wireless carrier must choose which interexchange carrier will be used for interLATA service so that the line can be *presubscribed* to the selected IC. Other ICs can still be accessed by prefixing the called number with the proper carrier code using the 10XXX format.

Dial lines enable the wireless carrier to access any valid telephone number. This includes NXX codes within the LATA as well as codes accessible through an IC's network, including international calls. The dial line also permits the wireless carrier to reach N11 codes (411, 911, etc.), and service access codes (SACs).

Direct Inward Dialing (DID) Connections

Unlike the dial line connection, *Direct Inward Dialing (DID)* connections are trunk-side end office connections. However, these two-wire circuits are limited to one-way, incoming (LEC-WC) service. In some cases, they are also considered to be Type 1 connections as discussed below.

Because they are trunk-side connections, DID connections employ different supervision and address pulsing signals than dial lines. Typically, DID connections use reverse battery supervision and either dial pulse or dual tone multifrequency (DTMF) address pulsing. There is no 20 Hz alerting signal (ringing) for these connections because the trunk is seized by the end office, and then the dialed digits are outpulsed by the end office. This outpulsing may be immediate or there may be a slight delay until the wireless carrier's switching machine indicates, via a *wink* signal, that it is ready to receive digits. These methods for outpulsing digit control are known as *immediate dial* or *wink start,* respectively. The more common method is wink start operation. DID connections are available from most switching machine types, including older electromechanical machines, and allow the wireless carrier to receive calls from other end offices as well as from interexchange carriers.

Type 1 Connections

This connection was the first type of connection offered to the cellular carriers. The Type 1 connection was developed because dial line and

DID connections were not suitable for use by cellular carriers. While sometimes referred to as a PBX-like connection, the only commonality Type 1 has with a typical PBX trunk is that both are served from an end office. Unlike the PBX trunk which is simply a dial line connection, the Type 1 connection is a trunk-side connection to an end office that uses trunk-side signaling protocols in conjunction with a feature generically referred to as *trunk with line treatment (TWLT).*

Basically, the TWLT feature allows the end office to combine some line-side and trunk-side features. For example, while trunk-side signaling protocols are used, the calls are recorded for billing purposes as if they were made by a line-side connection. TWLT also permits the trunk group to be presubscribed to an IC, similar to a dial line connection. However, unlike dial line connections, the use of TWLT enables the end office switch to return answer supervision to the wireless carrier.

Two-way trunks are always four-wire circuits, meaning they have separate transmit and receive paths, and almost always use multifrequency (MF) address pulsing and E&M supervision. The address pulsing normally uses wink-start control. One-way Type 1 connections can be provided on a two-wire basis using E&M supervision or reverse battery like the DID connection.

Using Type 1 connections, the wireless carrier can access any valid telephone number. This includes NXX codes within the LATA as well as codes that are accessible through an IC's network, including international calls. The Type 1 connection also permits the wireless carrier to reach Directory Assistance, N11 codes (411, 911, etc.), and service access codes (SACs).

Type 2A Connections

Type 2A connections allow the wireless carrier to connect to the PSTN like any other end office. These connections may be just to an access tandem, or a combination of connections to both access and local tandems. The latter may occur if the local exchange carrier uses separate switches for local and toll type traffic. However, no connection exists between a local tandem and an access tandem in this configuration, so overflow routing is not possible.

Type 2A connections are true trunk-side connections that employ trunk-side signaling protocols. Typically, they are two-way connections that are four-wire circuits using E&M supervision with multifrequency (MF) address pulsing. The address pulsing is almost always under wink-start control.

Since the wireless carrier's switch functions like an end office with a Type 2A connection, the local exchange carrier does not provide presubscription to an IC. Selection of an IC depends upon whether the wireless carrier offers equal access to its customers. If the wireless carrier offers equal access, the IC is selected by its customer and the call is routed to the proper IC by digits contained in the call protocol forwarded by the wireless carrier to the access tandem. If the wireless carrier does not offer equal access, the call may be routed to an IC selected by the wireless carrier, not its customer, based on digits forwarded to the access tandem.

If Type 2A connections to a local tandem are used, calls are restricted to valid NXX codes within end offices subtending the local tandem. Calls to or from a location outside of this calling scope must be routed to the access tandem. Therefore, a Type 2A connection to a local tandem can never be furnished without an accompanying connection to an access tandem as well.

There are some restrictions on the kinds of calls that can utilized with a Type 2A connection. Presently, access to Operator Services provided by the local exchange carrier and N11 codes (411, 911, etc.) is not permitted. Some local exchange carriers also restrict 800 or 900 traffic from the Type 2A. Therefore, at the present time, a Type 1 connection, or its equivalent, must be used in conjunction with the Type 2A to provide access that is not possible with the Type 2A alone.

Type 2B Connections

In the traditional local exchange carrier network, direct connections are used between end offices that send large amounts of traffic between each other. These trunks are established when it becomes more efficient and economical to employ direct trunks rather than routing all of the traffic through the tandem office. With direct trunks, the tandem is used only for overflow purposes. A similar arrangement is available to wireless carriers with the Type 2B connection, which is a trunk-side connection to an end office that functions like a high-usage trunk and is intended to be used in conjunction with the Type 2A connection to a tandem office. The first-choice route is the Type 2B with overflow through the Type 2A connection.

Because of its intended use, there are restrictions on the Type 2B connection. It can access only valid NXX codes served by the office providing the Type 2B connection. It cannot be used to reach Operator Services, N11 codes, service access codes, or trunk-side connections used to access an IC.

Type 2B trunks are not presubscribed to an IC for two reasons. First, access to an IC is not permitted. Second, since the wireless carrier's switch functions like an end office, any presubscription feature, if ever needed, would have to be provided by the wireless carrier.

Because they are like high-usage trunks, Type 2B connections are almost always four-wire, two-way connections that use E&M supervision and multifrequency (MF) address pulsing. As with other trunk-side connections, wink start control is used almost exclusively.

Private-line Connections

In order to connect the wireless carrier's switch or terminal to a radio site, or to connect two radio sites together, wireless carriers often use private-line connections from the local exchange carrier. These private-lines may be two-wire or four-wire analog circuits, high-capacity digital circuits at *DS1* or *DS3* rates, or special wideband analog circuits requiring the use of fiber optic facilities. These links are used to carry voice traffic or signaling data.

Interfaces Used

For all interconnection types described above, the interface at the *point of termination (POT)* can be analog or digital. The type of interface used depends upon the wireless carrier's equipment, the circuit quantities needed, and the rates and conditions offered by the local exchange carrier.

Analog interfaces are quite common when small circuit quantities are required and older equipment exists, as well as with service to some isolated locations. These interfaces can be two-wire or four-wire.

Digital interfaces are very common with higher circuit quantities and newer technology. The digital rate at the interface can be DS1 (1.544 Mbps) or DS3 (45 Mbps).

GLARE RESOLUTION

It does not happen often, but occasionally two switching machines will simultaneously seize a trunk that connects both machines. This event, known as *glare,* occurs because of an unguarded interval between the trunk seizure at one end and the subsequent making-busy of the same trunk at the other end. When glare occurs, the switching machines must have instructions for resolving this condition. This instruction is commonly known as a *glare bit,* which identifies which office is to retain control of the trunk and which releases the connection.

Line-side connections use different signaling sequences than trunk-side so glare resolution is only required for trunk-side connections. Obviously, glare resolution is employed only for two-way trunk-side connections because glare cannot occur with one-way trunks.

When two-way trunks are used, an agreement should be reached before the trunk is placed in service as to which switching machine will yield in the event of a glare condition. Traditionally, the lower ranked machine, or office in the switching hierarchy, will yield. Thus, end offices generally yield to tandem offices and tandem offices yield to IC offices. If the offices are of equal rank, glare resolution is often determined by the low alpha characters of the Common Language Location Identification (CLLI™) code, which is an 11-character code describing the location of the switching machine. However, these conventions are not entirely

Table 2.7
Call Progress Tones/Recorded Announcement Requirements

TONE ANNOUNCEMENT	INDICATION METHOD	PURPOSE	PROVIDED BY TRUNK CONN.	PROVIDED BY LINE CONN.
Audible Ringing	2 sec. on/4 off Ringing cycle	Notify Calling Party Of Ringing	WC	LEC
Busy	60 ipm audible signal	Called Party Off Hook	WC	LEC
Fast Busy	120 ipm audible signal	Network Trunks Not Available	WC/IC	LEC/IC
Vacant Number	Recorded Announcement	Number Dialed Not In Service	WC	LEC
Vacant NPA Code	Recorded Announcement	NPA or SAC Dialed Not In Service	WC/IC	LEC/IC

rigid and other resolutions can be devised. The important thing is to have an agreement in place before the trunks are placed in service.

For Type 2A connections, typically the wireless carrier's switching machine will yield since it is the equivalent of an end office and the tandem is considered to be the higher ranked office. The same is generally true of Type 1 connections. Type 2B connections would generally use the low alpha characters of the CLLI code convention.

CALL PROGRESS TONES AND RECORDED ANNOUNCEMENTS

Call progress tones are used to inform the calling party of the status while a call attempt is in progress. For example, the calling party will hear audible ringing as 20 Hz ringing to the called party begins or a line busy indication if the line is in use. For calls that terminate in a local exchange carrier's end office, call progress tones are provided by the local exchange carrier. Likewise, for calls terminating on a wireless carrier's network, the tones are provided by the wireless carrier.

Recorded announcements inform the calling party of special conditions. For instance, if a call is placed to an unassigned number in an end office, a recorded announcement provided by the local exchange carrier indicates that a vacant code has been reached. When the wireless carrier has the responsibility for assigning these numbers, it also has the responsibility of providing any necessary recorded announcements.

Special information tones are used to precede some announcements in order to get the attention of the calling party and allow automatic assessment of network performance. For example, a three-tone signal precedes a vacant code announcement. The responsibility for these special information tones is the same as that for the recorded announcements. Table 2.7 lists the various types of call progress tones and recorded announcements and their characteristics.

3 INTER-CONNECTION: OPTIONAL SERVICES

Like any other business, wireless carriers are anxious to improve service to their customers and reduce the cost of providing those services. Two options, *Calling Party Pays (CPP)* and *Wide Area Calling Plans (WACP),* are often requested from the LECs to improve service for the wireless carrier's subscribers. Two others, *Selective Class of Call Screening (SCCS)* and *Toll Billing Exception (TBE),* are useful in reducing costs by improving fraud control.

CALLING PARTY PAYS

Calling Party Pays (CPP) is a service whereby the calling party pays for the airtime of the wireless subscriber when a call is made to that subscriber. While it usually pertains to the landline caller (the LEC's subscriber), it is also possible when one wireless subscriber calls another wireless subscriber. When the latter case occurs, these wireless subscribers may or may not be customers of the same wireless carrier.

Many wireless carriers want this service because it comports with the concept of the calling party paying for any charges associated with that particular call. With the exception of special instances, like calls to "800" numbers, this concept is universally accepted in the United States for most landline-landline calls. For example, if you call Aunt Maude, you will normally pay all of the toll or message charges if the call is from your landline phone to Aunt Maude's landline phone. However, if you have a landline phone and Aunt Maude subscribes to a wireless carrier, you will usually pay the landline charges while she will pay for the airtime.

In Europe, the concept of the calling party paying for the airtime for wireless subscribers is widely accepted. Compensation between the landline (or LEC) and wireless carrier is accomplished in much the same manner as it is between two landline carriers. An agreement exists on the charges each will pay and it is generally based on the amount of traffic between the two systems. Each call is recorded for billing purposes. The inter-company agreements generally contain provisions for calls between nations when it is not possible for the receiving country to obtain all of the billing data when the call is placed. No special dialing procedure is necessary; the call is placed just like any other call.

In the United States, implementation of Calling Party Pays has been more difficult because there are many more carriers involved, and there are regulatory concerns as well. While the service has been implemented in a few places, widespread implementation has been slowed while issues such as customer notification, "revenue leakage," and NXX utilization are discussed. Consequently, in locations where Calling Party Pays is provided, the dialing arrangement is different from normal "local" calls, and the wireless carrier absorbs the cost of a number of calls that are excluded from the service.

Where Calling Party Pays has been implemented in the United States, it has required the assignment of a complete NXX code for this purpose. The wireless carrier offers its customers the choice of paying for the airtime or not. If the customer elects to receive the Calling Party Pays option, he is assigned a telephone number from the NXX which the wireless carrier has reserved specifically for this service.

One of the regulatory concerns is advising the landline customer that extra charges will be incurred for these calls. Consequently, where Calling Party Pays is currently implemented, the calling party must prefix the dialed digits with the digit "1." In most locations, this prefix is presently used to signify a toll call so the landline user expects to pay an additional charge.

However, not all LECs require the use of the "1+" prefix for toll calls today because in some cases it is used only if the call is to another NPA code. Moreover, with the introduction of interchangeable central office codes, "1+" dialing does not always indicate a toll call because it could be required to complete a call within the same local calling area.

Another solution is to use a recorded announcement, similar to the arrangement used for many calls to numbers within the 900 service access code. The announcement would alert the calling party to an extra charge.

From a wireless carrier's perspective, neither notification solution is desirable because they would prefer to have the calls completed using the same dialing procedures that are used for landline calls.

While the wireless carrier would like to have the landline customer pay all of the airtime charges for calls to a wireless subscriber who has elected the CPP option, there are a number of instances where this is not possible. The service area of the wireless carrier often encompasses several LECs, and in order for CPP to be effective, agreements must be reached with all of the LECs. There are other situations where it is currently not possible to charge the calling party for the airtime. These include calls from Wide Area Telephone Service (WATS) lines and coin phones plus interLATA calls involving an IC. All of these circumstances create "revenue leaks" that are usually absorbed by the wireless carrier.

WIDE AREA CALLING PLANS

Wireless carriers consider their service area to be their local calling area. But, as illustrated in Figure 3.1, these service areas often overlap the local exchange carrier's exchange boundaries, so calls from a landline customer to a wireless subscriber are not always "local." Because the wireless carriers believe that "toll" charges may discourage landline customers from calling a wireless subscriber, the wireless carriers often request the LECs to provide alternatives that eliminate toll charges to the landline user.

A traditional approach to this problem has been the use of numbers within the 800 service access code. There are 16 specific codes available for wireless carriers who are classified as *Radio Common Carriers (RCCs)* to use for intrastate service. However, in many states, the demand for these numbers is far greater than the supply provided by the 16 available codes.

Another alternative would be the assignment of a unique service access code to the wireless industry. However, there are limited codes available and, presently, there is no industry consensus regarding this solution.

A third solution is for the LEC to offer a Wide Area Calling Plan (WACP) that allows the wireless carrier, rather than the landline customer, to pay for calls made by a landline customer to the wireless subscriber. Ideally, the call can be dialed by the landline customer using the same procedure as for any call in the local calling area, but this is not possible in all circumstances.

Some local exchange carriers offer Wide Area Calling Plans, but the variations are numerous. Some plans are limited to specific interconnection arrangements (like Type 2) while others are restricted to particular types of wireless carriers (cellular carriers only). Some LECs define the specific coverage area while others allow the wireless carrier to select the coverage area.

As with Calling Party Pays, the wireless carrier often has to negotiate these arrangements with several LECs because its wireless service area is not confined to the territory of a single LEC.

Because of billing system limitations, a NXX code that is dedicated either to a specific wireless carrier, or to this specific service, is required for any of these variations. Calls to this dedicated NXX are recorded at the end office serving the landline user making the call. Typically, charges are assessed to the wireless carrier based on the accumulated usage for all calls that were made to numbers within that dedicated NXX code during the billing period. The usage charges may be adjusted to account for those calls that would be considered "local" without this plan.

Wide Area Calling
FIGURE 3.1

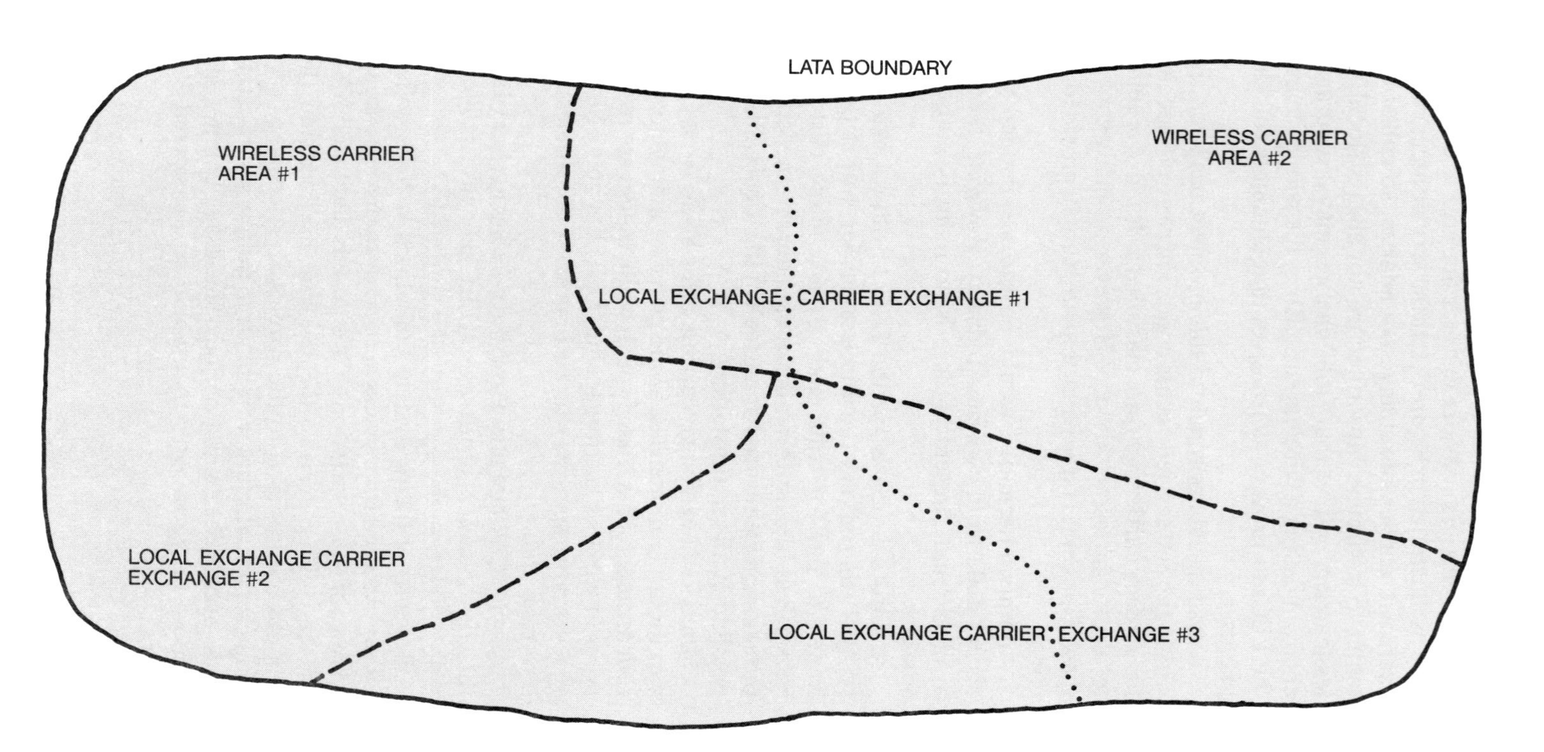

Wide Area Calling Plans compensate for the fact that the service areas of the wireless carriers overlap the exchange boundaries of the local exchange carriers by allowing the wireless carriers to pay usage charges for land-to-mobile calls rather than the landline subscriber paying toll charges to call between exchanges.

TOLL FRAUD REDUCTION

There are two options that wireless carriers often order from the LECs that are not fundamental requirements for any specific interconnection arrangement (dial line, Type 1, 2A, etc.). Nonetheless, these options provide a valuable service by reducing, but not totally preventing, fraud for the wireless carriers. While they may have other names, they are generally referred to as *Toll Billing Exception (TBE)* and *Selective Class of Call Screening (SCCS)*.

Toll Billing Exception

Toll Billing Exception (TBE) is designed to eliminate third-number billing to a telephone number assigned to a wireless carrier. When the wireless carrier requests the TBE feature, the telephone numbers assigned to the wireless carrier are entered into a database that can be accessed by LEC operators as well as by some IC and other LEC operators. Presently, the Billing Validation Application (BVA) database owned by American Telephone & Telegraph (AT&T) performs this function, but most of the larger LECs will soon have their own Line Information Database (LIDB), which will replace BVA. When a caller attempts to place a call, and requests that the charges be applied against a telephone number assigned to the wireless carrier, the database alerts the operator that billing to that number is not allowed.

TBE can be used with any interconnection arrangement and is not limited by switch type, since it is only a database entry. However, its effectiveness is restricted because not all of the companies providing operator services subscribe to these databases. Consequently, most LECs confine their liability for any fraud that does occur to calls that are handled by their own operator services group.

SELECTIVE CLASS OF CALL SCREENING

Selective Class of Call Screening (SCCS) is a feature that can be used with dial line or Type 1 connections to restrict wireless-originated calls that require operator assistance to collect calls, calling card, or third number billing only. Unlike TBE, this feature is not a database entry but is a function of the switching machine. While a number of switching machines can provide SCCS when it is associated with a dial line connection, it is not widely available when Type 1 connections are used.

Without the SCCS option, calls requiring operator intervention are billed back to the wireless carrier rather than to the wireless subscriber. Consequently, many wireless carriers insist on the SCCS feature if it is available from the LEC.

4 TWO-WAY MOBILE SYSTEMS

SERVICE EVOLUTION

Almost from the time the automobile began to be mass-produced, efforts were made to incorporate a radio that would allow the occupant to communicate with others while driving. These early efforts were aimed primarily at public safety agencies and did not include the ability to interconnect with the traditional landline telephone network. Very early systems were also limited to one-way transmission, but by the mid-1930s, had expanded to two-way communication. Two-way mobile systems have evolved into two basic categories: dispatch-type service and *public land mobile service (PLMS).*

Dispatch service is designed for communication between vehicles, or between a vehicle and a dispatcher. It has the capability to be interconnected with the *Public Switched Telephone Network (PSTN)* but its major function is to facilitate communications within the vehicular operation.

Public land mobile service is designed for communication similar to that of a landline telephone. It is fully interconnected with the PSTN and its purpose is to facilitate communications between mobile and landline subscribers. It is this type of service that is described in this chapter.

Although it was limited to emergency vehicles, a system was provided in 1940 by New York Telephone in New York City that was capable of operating on an integrated basis with the landline telephone network. Commercial service had to wait until 1946 when it was introduced in St. Louis by Southwestern Bell. It had only a single channel, needed the use of an operator, and the mobile phone required push-to-talk operation. This was necessary because the user could either listen or talk, but could not do both simultaneously as with a landline telephone. This type of operation is known as *half-duplex* operation.

As systems grew, channels were added. Initially, the mobile set was assigned to a particular channel. This type of operation is referred to as a *non-trunked* system. Then, to improve channel utilization, the mobile user was given the ability to use any channel. These arrangements were called *trunked* systems. Early *trunked* systems required the mobile user to search manually for a vacant channel using a channel selector. Eventually, this process was automated so that the set automatically scanned for a vacant channel. Half-duplex operation remained the standard until the 1960s when *Improved Mobile Telephone Service (IMTS)* was introduced by the Bell System. IMTS allowed two-way simultaneous conversations, or *full-duplex* operation, just like a landline telephone. In addi-

Typical 2-Way Mobile System
FIGURE 4.1

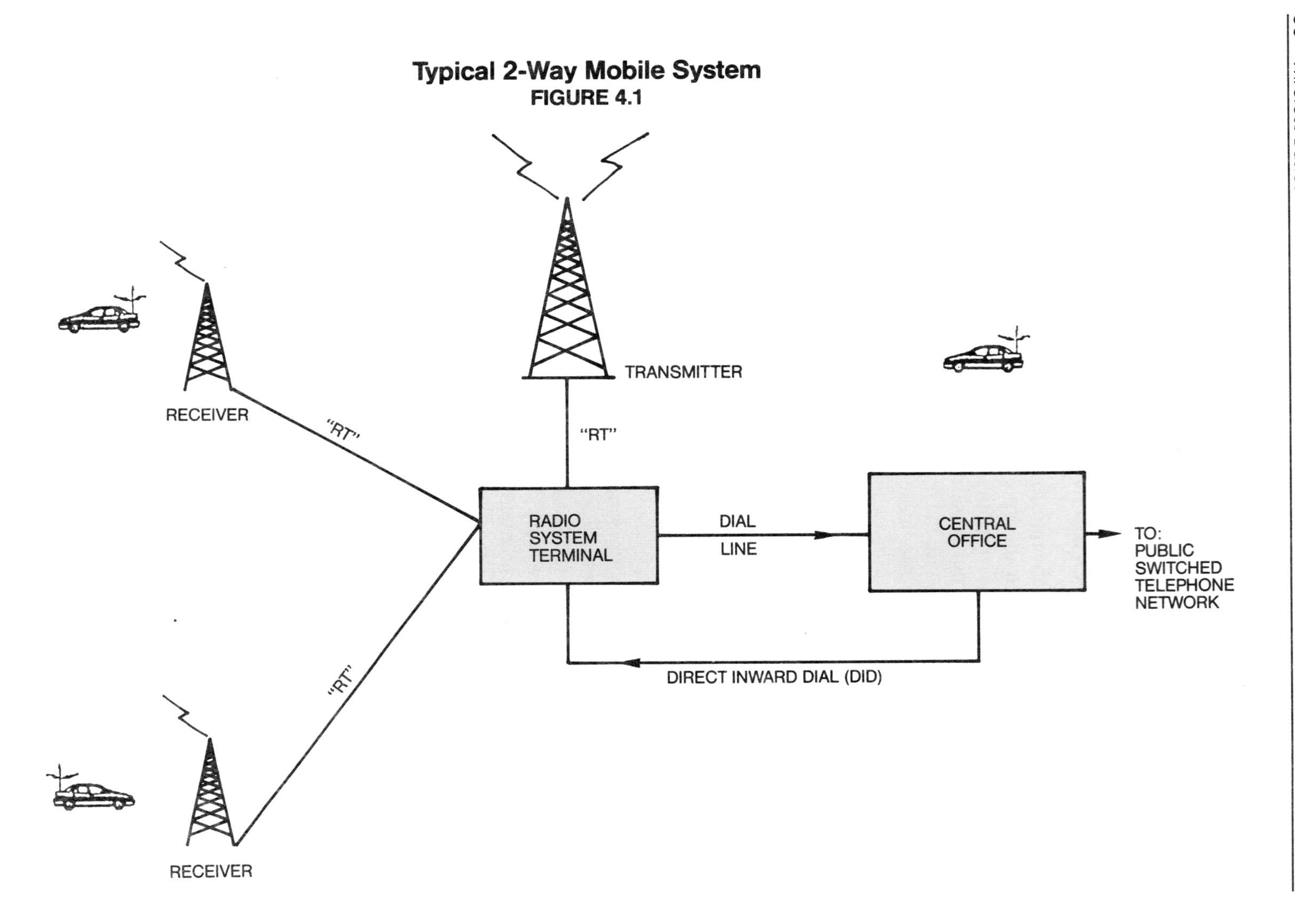

tion, IMTS eliminated the need for operator intervention by allowing the mobile user to dial the number instead of placing a call through an operator. Interestingly, while automatic dialing did not become widely available until IMTS was introduced, some non-Bell *local exchange carriers (LECs)* had equipment in the late 1940s that provided dial capability with half-duplex operation.

Providers of two-way mobile service were not restricted to telephone companies because this was one of the first areas where the *Federal Communications Commission (FCC)* introduced competition. In 1949, the FCC authorized non-telephone company entities, known as *Radio Common Carriers (RCCs),* to provide two-way mobile service. In fact, for some time, the RCCs actually had the largest share of the two-way mobile market. While later popularized with cellular service, this decision introduced the terms *wireline* and *non-wireline.*

RADIO SPECTRUM REQUIREMENTS

Two-way mobile service dedicated to the PLMS category operates in two frequency bands authorized by the FCC. These bands are in the *Very High Frequency (VHF)* and *Ultra High Frequency (UHF)* bands. Each utilizes two sets of frequencies within those bands in order to provide the full-duplex service. For base-to-mobile operation, the VHF service uses frequencies in the 152 MHz, range while the 157-158 MHz frequencies are used for mobile-to-base transmission. Mobile services utilizing the UHF band use frequencies in the 454 MHz range for base-to-mobile and 459 MHz for mobile-to-base. Very early PLMS systems used frequencies in the 35 MHz range but this spectrum is now limited to dispatch or paging services.

Because of the lower frequency, the VHF systems usually provide greater range and are more dependable than the UHF systems. This is particularly true in remote areas where the VHF signal travels further and is much more resistant to signal loss due to terrain features.

Some two-way mobile service that is used primarily for dispatch type operation, but has interconnection to the PSTN, is provided by *Specialized Mobile Radio (SMR)* systems. These SMR systems use frequencies in the 800 MHz and 900 MHz bands and are presently expanding rapidly.

BASIC SYSTEM ARCHITECTURE

As illustrated in Figure 4.1, conventional two-way mobile systems, like IMTS, employ a single transmitter that emits a powerful signal strength of up to 500 watts *effective radiated power (ERP).* In order to cover as large an area as possible, the transmitter is usually located on the highest point available and operated at the highest permissible power level for that height. In order to minimize interference with other systems, transmitters for systems utilizing the same frequencies must be separated by about 70 miles. Consequently, not all areas are effectively served by two-way mobile service.

Mobile units transmit at a much lower power level, usually about 25 watts, and obviously the mobile antenna is normally at a much lower level relative to the transmitter antenna. In order for the range of a two-way mobile system to be the same in both directions of transmission, these systems often employ remote sites that contain receivers. These remote sites detect the mobile unit's transmission and in turn forward it back to the system controller for processing.

The system controller allocates available channels to mobile users initiating or receiving calls. Many systems employ a comparator circuit in the system controller that constantly monitors the signal quality of the remote receivers and chooses the signal from the receiver site with the best signal-to-noise ratio. The controller may also change receiver sites during the course of the call if the signal quality decreases below specified parameters. In addition to controlling the transmitters and receivers, the system controller determines if the mobile user is authorized to use the system and records the call for billing purposes.

Channel Bandwidth and System Capacity

Although the early systems required up to 120 kHz of bandwidth for each channel, gradual improvements have been made so that present versions are allotted 30 kHz per channel for VHF systems and 25 kHz per channel for UHF systems. IMTS has a designed maximum of 11 VHF or 12 UHF channels per system. Equipment used by the RCCs has similar channel bandwidth requirements. It shares the spectrum authorized by the FCC of 8 VHF and 14 UHF channels for the RCCs.

Due to FCC rules, only 8 channels can be utilized in any one location. However, for a large metropolitan area like Los Angeles or New York, all 18 VHF or 26 UHF channels are effectively utilized because they may be divided amongst cities that comprise the metropolitan region. Furthermore, some cities have both VHF and UHF systems while others have only one. Many small cities have only one single channel to serve their area.

RCC systems in several metropolitan areas also use frequencies in the lower part of the UHF television band. For example, in Miami 12 frequencies in the 470-473 MHz region (TV Channel 14) are authorized for mobile use.

GRADE OF SERVICE

Landline telephone networks are usually engineered so that about 99 of every 100 calls during the busiest time will be completed. This is known as a B.01 grade of service. Because of the limited number of radio channels and, prior to cellular service, the high demand for conventional mobile services, two-way mobile systems like IMTS are engineered for a B.50 grade of service. This means that only 50 of every 100 calls are completed during the busy period. A typical multichannel system was designed to have about 100 mobiles per channel. However, it was not unusual for the actual performance of the system to be far worse than B.50 because service providers often added more customers than the system was designed for in an effort to meet demand.

NUMBERING REQUIREMENTS FOR TWO-WAY MOBILE SERVICE

Each mobile unit using a conventional two-way mobile service like IMTS is assigned a telephone number that conforms to the *North American Numbering Plan (NANP).* The carrier providing the service obtains blocks of telephone numbers from the *Local Exchange Carrier (LEC)* and assigns numbers to its mobile subscribers. This number becomes part of the *mobile identification number (MIN),* but, particularly for IMTS systems, the numbers had to be carefully coordinated to avoid a conflict within a particular *Numbering Plan Area (NPA).* While the use of numbers from a dedicated NXX code is possible, the small capacity of the two-way mobile systems does not normally warrant the assignment of a dedicated NXX code.

INTERCONNECTION TYPES

Conventional two-way mobile systems are designed to operate using *direct inward dialing (DID)* trunks for incoming (land-to-mobile) calls and dial lines for outgoing (mobile-to-land) calls. While other interconnection arrangements are possible, they would require new equipment design changes that may not be economically practical. In order to connect the system controller to the remote receiver sites, the two-way mobile carrier will often use private line links supplied by the LEC. These links, often known as *radio telephone (RT)* circuits, are voice-grade circuits that have no signaling capability provided by the LEC. Digital links, such as DS1 rate facilities, could be used, but the circuit requirements for these private line facilities are almost always too small to justify digital facilities.

INTERFACE REQUIREMENTS

Because of the small quantities required, and the design of the equipment used for conventional two-way mobile service, two-wire analog interfaces are almost exclusively used by carriers providing this service.

CALL PROCESSING

Two-way mobile systems must be able to identify the user in order to process the call. In earlier systems, this was done solely with the use of an operator. Later systems have the ability to identify the subscriber based on the *mobile identification number (MIN),* which is actually the NPA plus the last four digits of the assigned telephone number. Many RCCs used unique system identification numbers that helped identify the user to their systems. This information was "hardwired" into the original mobile sets, as opposed to later technologies like cellular that encode similar information using electronic *read-only memories (ROMs).*

The information below pertains to IMTS-type systems used by telephone companies because their operation is standardized. RCCs employ equipment from a number of vendors, not all manufactured to the same standard, so it is difficult to provide a universal description. The dialing sequences are similar between the IMTS-type systems and those used by RCCs but the encoded information and processing efforts may differ. This is also why roaming between RCC systems is almost non-existent.

Most modern two-way mobile sets have three modes of operation that are manually selected by the mobile user. The selected mode is indicated by lamps on the set. These are *home, roam,* and *manual.* In dial systems, the *home* mode is used when the set is operating in its normal location. *Roam* is selected if the mobile is away from its home area, while *manual* is available for any area still requiring the use of an operator.

Mobile-To-Land Calls—Home System

When a two-way mobile subscriber obtains service from a LEC, the mobile unit is wired so that it will transmit its MIN to the system controller after it has seized a channel. In addition, the mobile set is arranged so that it searches over only the channels used by that particular system. After seizing a vacant channel and transmitting its MIN, the system controller returns a dial tone indication to the mobile user after determining the validity of the user. The mobile subscriber then dials the digits which are transmitted to the system controller as the mobile user dials them. The system controller seizes a dial line circuit, outpulses the digits, and the call is processed through the PSTN like any other call.

Manual system operation varies somewhat depending on the carrier providing the service. Generally, the mobile user depresses the push-to-talk button on the set to obtain a channel to an operator. The operator then obtains the billing information and then dials the number the mobile user wishes to call.

Land-To-Mobile Calls—Home System

With a normal two-way mobile system, the landline caller dials the telephone number assigned to the mobile unit. The call is directed to the LEC end office providing the DID connection to the two-way mobile system, and the digits of the mobile unit are outpulsed over the DID trunk. Only the last four digits of the mobile unit are outpulsed. The system controller verifies the digits, seizes the next vacant channel, and broadcasts the digits of the mobile unit. The mobile set, which constantly resets itself to the next vacant channel, receives the digits and generates a ringing signal to alert the mobile user. When the call is answered by the mobile user, the voice channel is cut through and the conversation can begin.

With a manual system, incoming calls to a mobile unit are routed to a mobile operator. The mobile operator then manually seizes the next vacant trunk, and initiates a broadcast of the mobile unit's telephone number. From that point on, the sequence is the same as that of an automatic dial system.

911 Traffic From A Two-Way Mobile Telephone

Emergency calls using the 911 dialing pattern are possible with a two-way mobile phone, but the carrier must first make arrangements with the local public safety agency before such calls can be forwarded to the PSTN. Sometimes the public safety agency will want these calls translated to a special seven-digit telephone number so that they are routed to a special position at the agency. If 911 calls are acceptable, the call is processed in the normal mobile-to-land sequence just described.

ROAMING WITH A TWO-WAY MOBILE TELEPHONE

Roaming is not normally possible with RCC systems because of the lack of equipment standardization. For IMTS-type systems, roaming is possible in many locations. In fact, conventional IMTS-type two-way mobile systems are still the only way mobile users can obtain service outside their home areas in many parts of North America. Roaming is possible with both VHF and UHF systems although there are more VHF systems in operation.

Mobile-To-Land Calls—Roaming

When a mobile user travels away from the normal service area, he/she manually changes the mode of operation on the set by depressing the *roam* button. This enables the set to scan all of the channels on its frequency band. The method used to place a call depends on the *visited* system, and directories are published by several firms to provide exact instructions. In some cases, the call is processed automatically as if the mobile user had placed the call in the "home" system. In other instances, the mobile user must obtain service through an operator by using the push-to-talk key on the mobile set.

Land-To-Mobile Calls—Roaming

Landline users may contact a mobile user when they are located in another area, but the procedure used varies depending upon the location. In either case, it is not as easy as making a call to a mobile in its home location and requires some degree of sophistication on the part of the landline caller. The landline caller must know the city in which the mobile unit is located and whether to place the call through a mobile operator or the "roamer" port for that city.

If a mobile operator is required to complete the call, the landline caller asks the LEC or *Interexchange Carrier (IC),* depending on whether the call must be transported outside the *Local Access Transport Area (LATA),* to connect with the mobile operator for the location he/she is trying to reach. Upon reaching the mobile operator in that area, the landline caller provides the telephone number of the roaming mobile unit he/she is trying to contact. The mobile operator then broadcasts the number of the mobile unit, and the call is processed like any other land-to-mobile call.

Many systems employ a roamer port that permits a landline caller to reach a mobile unit without the intervention of an operator. The landline caller dials the roamer port number of the system where the mobile unit is visiting. The roamer port number is simply a normal telephone number that terminates in the system controller of the carrier. The call is routed through the PSTN. When it reaches the end office serving the visited system, the digits of the roamer port are outpulsed over the DID trunk that connects the system controller to the end office. A dial tone is returned to the landline caller prompting him/her to dial the number of the mobile unit using a push-button phone. The system controller then broadcasts the digits of the mobile unit, and the call is processed in the normal manner.

5 CELLULAR MOBILE SYSTEMS

SERVICE EVOLUTION

Commercial public mobile service began in the United States in 1946. It had hardly begun when its inherent capacity limitations were recognized. In 1947, Bell Telephone Laboratories began exploring a concept that would reuse frequencies by utilizing small cells with low-powered mobile units. These cells could be linked together using a computer to permit mobility while simultaneously greatly increasing the number of subscribers that could be served by the system. However, the necessary computer technology did not become available until the mid-1960s when electronic switching systems were developed.

In spite of the fact that the cellular concept was developed in the United States, the first commercial cellular systems were installed in Japan in 1978, followed by the Scandinavian system in 1981. Largely due to regulatory problems, commercial cellular service did not begin in the United States until 1983 when the then-experimental Chicago system became operational.

While cellular systems have been implemented in over 70 countries, there is no worldwide standard for these systems. The most popular system is the based on the United States version called *AMPS (Advanced Mobile Phone System),* which is the acronym for original name for the service when it was developed by Bell Laboratories. Other versions include *NMT (Nordic Mobile Telephone), TACS (Total Access Communications System),* and *GSM* (originally known as *Groupe Speciale Mobile* but now also stands for *Global Systems for Mobile).* These systems have different operating characteristics. Even when the same technology is used by various countries, e.g., AMPS, the systems often utilize different frequency bands.

REGULATION AND COMPETITION

As further explained in Chapter 9, the *Federal Communications Commission (FCC)* issues two cellular licenses for each defined geographical area in order to promote competition. These geographical areas are known as a *Metropolitan Statistical Area (MSA)* or *Rural Service Area (RSA).* One of these licenses belongs to a subsidiary of a company that provides landline telephone service in the given geographical area and is known as the "wireline" licensee, or "B" band. The other belongs to a company not associated with the landline telephone company in that area and is called the "non-wireline" licensee, or "A" band. Due to mergers and acquisitions in recent years, this distinction has become very confusing because "wireline" companies often operate as "non-wireline" carriers outside of the franchised areas of their landline telephone companies.

A cellular licensee, often called a *Cellular Mobile Carrier (CMC)*, may serve all or just a portion of a designated MSA or RSA. Usually most of a MSA is served while large portions of a RSA may be left unserved. The actual service area is defined by the CMC and is called the *Cellular Geographic Service Area (CGSA)*.

Debate continues as to whether the FCC's effort to promote competition has been successful because there is often little difference in the prices charged by the wireline and non-wireline licensees or in the quality of service. Nonetheless, the overall price of cellular service has continued to drop and the number of subscribers has increased dramatically since service began in 1983. As of January, 1992, there are approximately 7.5 million cellular subscribers in the United States and the growth rate is still almost 40% per year. It took the cellular industry only 4 years to acquire 1 million subscribers which is remarkable when you consider it took 20 years for the telephone industry and 11 years for television to achieve that same level of acceptance.

RADIO SPECTRUM REQUIREMENTS

Each cellular licensee in a MSA or RSA is authorized by the FCC to use 25 MHz of spectrum within the 800 MHz range. Originally the FCC allocated 20 MHz to each carrier which was obtained by deleting UHF television channels 70-83 because they were lightly used. An additional 5 MHz per carrier was authorized in 1985 due to growing capacity problems in some areas. This 25 MHz of spectrum is divided for use by *base* and *mobile* stations within each system so that different frequencies are used for base to mobile and mobile to base. This type of operation is called *full duplex* and permits simultaneous conversation or transmission. It must be remembered that two channels are required for each conversation in order to operate on a full-duplex basis.

The existing cellular systems in the United States are based on an analog technology that employs *frequency modulation (FM)*. With this technology, each channel occupies 30 kHz of bandwidth yielding a total of 833 channels, or 416 channel pairs, for each cellular operator. Of these 416 channel pairs, 21 are dedicated to control functions for call set-up and handoff purposes. This leaves 395 channel pairs available for voice services.

Digital coding schemes are just now being introduced. Through the efforts of the *Cellular Telecommunications Industry Association (CTIA)*, an announcement was made in 1989 by the cellular industry it would adopt the *Time Division Multiple Access (TDMA)* multiplexing technique as the new digital standard. Tests have continued since that time and cellular sets that will accept either TDMA or existing analog signals will be marketed beginning in 1992. Other techniques, such as *Code*

Division Multiple Access (CDMA), are being explored but TDMA is the industry accepted digital standard.

BASIC SYSTEM ARCHITECTURE

Unlike two-way mobile systems that feature one large, powerful transmitter to cover a given geographical area, cellular service is based on the concept of having transceivers, i.e. a combination transmitter and receiver, located in small areas called cells. The number of transceivers in a given cell depends on the traffic that the cell is expected to handle. The cells can range in size from a radius of less than 1 mile to 25 miles or more with the range being determined by the transmitted signal power and height of the antenna.

By design, cellular systems operate at a lower signal power level than traditional two-way mobile systems. While the maximum power can be as high as 500 watts, it is generally around 50-100 watts for the base station. Mobile units designed for use in cars are capable of transmitting at 3 watts while portable, hand held, units have a maximum power of 600 milliwatts, or 6/10ths of a watt.

While cellular systems operate at a much lower power level, and the cells cover a much smaller geographical area than two-way mobile systems, the real advantage of cellular is due to the fact that a cellular system can reuse its frequencies in different cells. Consequently, the actual total number of radio channels far exceeds 833 due to this reuse ability. This reuse capability results from having the cells linked to a central switch that controls the assignment of the frequencies within a cell as well as the handoff capability that transfers the call from cell to cell as the mobile user moves through the service area.

Figure 5.1 illustrates a typical cellular system. In this diagram, the cell arrangements are shown in a hexagonal pattern. Cellular systems are engineered using this hexagonal arrangement in order to determine the frequency reuse pattern. This method is used to ensure enough separation exists between any two cells using the same frequencies so that interference does not occur. Typically, the system may be designed using a 4-cell, 7-cell, or 12-cell reuse pattern. If the world were flat and there were no obstructions, the radio waves could be confined to these hexagonal cells but in reality, the radio signals can, at best, only approximate the hexagonal shape. Special measures are needed for areas that are difficult to serve with normal antennas, like tunnels. Still, the use of hexagonal cells is a useful engineering tool.

CELLULAR SYSTEM GRADE OF SERVICE

Cellular systems are typically engineered for a B.01 or B.02 blocking probability, meaning 98 or 99 calls out of every hundred will complete during the busiest period of the day for the cellular system. This is another great improvement of cellular systems over the two-way mobile

Typical Cellular System

FIGURE 5.1

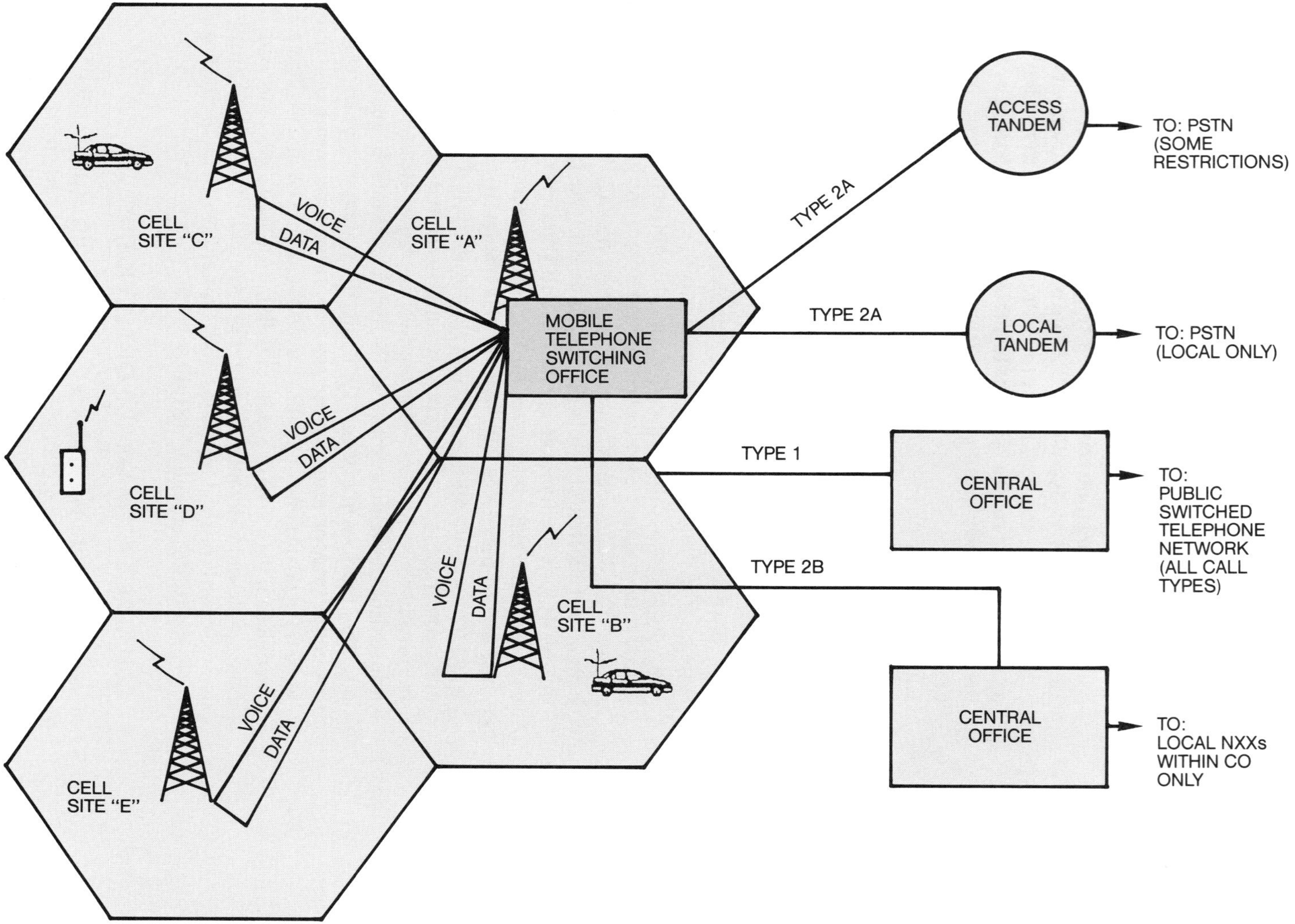

ACCESS TANDEM
TYPE 2A
TO: PSTN (SOME RESTRICTIONS)

LOCAL TANDEM
TYPE 2A
TO: PSTN (LOCAL ONLY)

CENTRAL OFFICE
TYPE 1
TO: PUBLIC SWITCHED TELEPHONE NETWORK (ALL CALL TYPES)

CENTRAL OFFICE
TYPE 2B
TO: LOCAL NXXs WITHIN CO ONLY

MOBILE TELEPHONE SWITCHING OFFICE

CELL SITE "A"

CELL SITE "B"
VOICE
DATA

CELL SITE "C"
VOICE
DATA

CELL SITE "D"
VOICE
DATA

CELL SITE "E"
VOICE
DATA

because it provides essentially the same blocking probability as a landline telephone system.

CELL SPLITTING AND SYSTEM CAPACITY

With the present cellular technology, a fully developed cellular system can conceivably support up to 500,000 mobile users. System capacity can be increased by increasing the total spectrum available, decreasing the size of the cells, or decreasing the bandwidth required for each channel.

Increasing Total Spectrum

As traffic grows in a cellular system, additional channels and cells are added until all of the available spectrum is in service. Metropolitan areas may use all of the spectrum but RSAs rarely need a full complement of channels. As explained in Chapter 9, only the Federal Communications Commission can authorize new spectrum for any radio service. Since spectrum is a very scarce resource, it appears that cellular operators will be restricted to their existing 25 MHz of spectrum for the present time.

Cell Splitting

In order to minimize interference, a certain distance must be maintained between cells using the same frequencies. However, this distance can be reduced without disturbing the cell reuse pattern. As the size of the cells are reduced, the same frequencies can be utilized in more cells, which in turn means more subscribers can be accommodated on the system. Particularly in congested areas, the cellular operator often splits an existing cell into two or more smaller cells. New transceivers are placed and the power of the transmitters reduced in order to confine the signals to the newly created cells. For example, a cell that originally had a radius of 8 miles could be split into four cells with each new cell having a 2 mile radius. For the existing analog systems, cell splitting is an effective way to increase system capacity although some practical limitations are reached. Suitable locations for cell sites becomes more difficult and the processing load on the switch rapidly increases because "handoffs" are more frequent.

Decreasing Channel Bandwidth

Existing systems require 30 kHz of bandwidth for each channel. It is possible to reduce the amount of channel bandwidth in order to increase system capacity. A technique called *Narrowband AMPS (N-AMPS)* has been introduced by Motorola permits an immediate 3-fold increase in capacity by using special coding techniques so that only 8 kHz of bandwidth is needed for each channel. N-AMPS offers an immediate, non-digital solution for capacity problems, and Motorola is producing cellular sets that can utilize both traditional and N-AMPS channels.

In 1992, dual mode sets that can use new digital standard (TDMA) or existing analog signals will become available. TDMA will also offer an immediate 3-fold increase (assuming all channels eventually use the TDMA signal) and a potential 10-fold increase as voice coding techniques improve.

Investigation continues into other multiplexing schemes, and it is estimated that potentially a 20-fold increase could be realized with other digital multiplexing schemes. These expansions will result from putting very high signal-processing power into the cellular sets via new integrated circuits.

NUMBERING REQUIREMENTS FOR CELLULAR SYSTEMS

Each cellular telephone set is assigned a telephone number that conforms to the North American Numbering Plan (NANP). This number is also known as the *Mobile Identification Number,* or *MIN.*

A CMC may utilize entire NXX codes or obtain blocks of numbers from a local exchange carrier (LEC) in order to assign the telephone numbers to its mobile users. Whether entire NXX codes or blocks of numbers are used depends on the connection types employed by the CMC and its total numbering requirements. Type 1 connections may use numbers from either a shared or a dedicated NXX code. Type 2 connections must use a dedicated code.

While dedicated NXX codes are often used with a Type 1 connection, the code itself resides in the LEC end office location. Thus, the *Vertical and Horizontal (V&H)* coordinates for codes associated with a Type 1 connection, as well as the *Common Language Location Identifier (CLLI™)* code, are those established for the LEC end office. The V&H coordinates define an unique geographical location and are used for rating purposes, as in the calculation of distance for toll charges. The CLLI code provides a name that is used by a number of operation systems, including those used for billing and routing.

With a Type 2 connection, a dedicated code is almost always needed. The V&H coordinates and a CLLI code are associated with the CMC's location instead of the LEC end office. Usually, but not always, this location is the CMC's switching center, which is called the *Mobile Telephone Switching Office (MTSO).* Because it is not always cost effective, the CMCs do not have a MTSO in every single service area. Frequently, their cell sites are connected to MTSOs in distant locations. For this reason, the *Point Of Termination (POT)* is used for the official location of the NXX code instead of the actual MTSO location. To use the MTSO location in these instances would be very misleading for call routing and billing purposes since it may well be physically located in another Numbering Plan Area (NPA) and state.

INTERCONNECTION TYPES

Three types of interconnection arrangements are currently available to connect a CMC's network to the LEC's network. These are known as Type 1, Type 2A and Type 2B. The actual interconnection arrangement used depends on the choice of the CMC as well as the technical capabilities of the LEC's switching machines.

In addition, the cellular system also requires links between its switching office, called a *Mobile Telephone Switching Office (MTSO),* and the various cell sites. These links provide both voice and signaling channels, each separate from the other. Often, the CMC will use its own microwave facilities for these links but private lines, particularly digital DS1 links, from the landline telephone company are also used.

INTERFACE REQUIREMENTS

Practically all of the cellular switches are digital and the majority of the facilities serving the CMC's location are digital. Therefore, regardless of the type of interconnection used, digital interfaces are usually less expensive and require less equipment from the LEC than analog facilities.

Because information about signal power is contained in the digital signal itself, the use of *Transmission Level Points (TLPs)* at each end of a trunk to describe its *Inserted Connection Loss (ICL)* is inappropriate because digital trunks have no ICL value. Instead, digital trunks are designed to be transparent, and their loss is a function of the decode level selected by the digital switch. This decode level refers to a 1004 Hz signal, known as a *Digital Reference Signal (DRS).* Generally, the DRS produces a level that is equivalent to a -6dB TLP after decoding.

Particularly with Type 1 connections, the POT interface at the CMC's location is digital, but the LEC switch is analog. However, as the LECs replace existing analog electronic switches with digital switches, the entire connection becomes digital. This is already the case with many Type 2A connections because most of the tandem switches are already digital machines.

Analog interfaces are available and are usually used when the number of trunks required is very small or when the CMC has a POT location in a remote area where digital facilities are not readily available. Because the transmission levels at the interface are usually +7 TLP (LEC Transmit) and -16 TLP (LEC Receive), some type of channel terminating equipment needs to be installed by the LEC at the POT location.

Digital Synchronization

When digital switches and/or digital facilities are interconnected, some means of synchronizing the clock rates must be used in order to avoid errors. The synchronization is obtained from the bit stream of a digital signal, such as a DS1 rate (1.544 mbps) channel, that is used to connect the two networks. An agreement must be reached between the LEC and the CMC regarding the primary source of timing.

CALL PROCESSING

In order to process calls in a cellular system, the location and identity of the mobile user must be known by the MTSO. This is true regardless of whether the mobile unit is operating in its "home" system or has traveled to a distant location and is considered to be "roaming." To accomplish this task, the mobile unit transmits certain information each time the power is turned on and again each time a call is originated. Specifically, this information consists of the mobile unit's *Mobile Identification Number (MIN)*, its *Electronic Serial Number (ESN)*, and its *System Identification (SID)*. This information is electronically encoded in the set itself and can be changed only by trained technicians.

Mobile-To-Land Calls—Home System

Dialing from a cellular phone is different from a landline telephone. Instead of receiving dial tone and then dialing the desired number, the cellular user first determines if service is available by verifying the presence of a "service available" indicator. This indicator is displayed when the mobile unit can receive a signal from one of the control channels in the cellular system. Assuming service is available, the mobile user enters the digits of the party they are calling and presses the "Send" button on their mobile unit. The mobile unit transmits these digits to the base station using the control channel. After ensuring the availability of a voice channel and trunk, the MTSO seizes a trunk, Type 1, 2A, or 2B depending upon the connection used and call type, and outpulses the required number of digits to the LEC office. When the call is answered, the MTSO connects the mobile to the voice channel so conversation can begin.

Land-To-Mobile Calls—Home System

To reach a mobile subscriber in a CMC system, the landline customer dials the 7 or 10-digit mobile number, depending on the local dialing plan. The call is directed to either the end office providing a Type 1 or Type 2B connection, or the tandem office if a Type 2A connection is employed. In either case, the 7-digit number of the mobile subscriber is outpulsed by the LEC office to the MTSO. If the MTSO has not recorded the presence of the mobile unit when the power was first turned on, the location of the mobile subscriber is determined by broadcasting a paging code that includes the mobile subscriber's MIN to all cell sites via a separate control channel. The cell site with the strongest signal strength from the responding sites is selected and a voice channel is assigned by the MTSO. A code is then outpulsed over the signaling channel which causes the mobile unit to initiate a ringing signal. When the mobile subscriber answers, the voice channel is connected so conversation can begin.

Mobile-To-Mobile Calls

Mobile subscribers can also reach other mobile subscribers, including those that operate on other cellular systems. If the call is to a mobile subscriber in the same cellular system, the MTSO sets up the call, via the control channels, between the cell sites serving the two mobile subscribers. If the call is to a mobile subscriber in another system, the call is normally processed like the mobile-to-land call described above. It is possible for direct trunks between the two MTSOs to be used to complete these calls.

Call Handoffs

As the mobile subscriber continues to move while the call is in progress, the original cell site is constantly monitoring the strength of the mobile's signal. When the signal begins to diminish, the cell site requests the MTSO to request signal measurements from adjacent cells to determine if the call should be "handed off" to an adjacent cell. Once this determination is made, the MTSO sends a message is sent to the mobile unit, via the control channel, advising the unit to change to a new frequency, and the call is transferred to the new cell site. This process takes less than one-half of a second and is imperceptible to the mobile subscriber.

Emergency Services Traffic From A Cellular Mobile Unit

In most cases, the CMC must make arrangements with the local public safety agency that is responsible for handling emergency calls before they begin forwarding such traffic to the LEC. Normally, the mobile user dials 911, but as noted in Chapter 2, 911 traffic can only be routed over Type 1 connections because it is presently restricted from Type 2A or Type 2B connections.

When a 911 call is forwarded by a CMC over a Type 1 connection, the *Automatic Number Identification (ANI)* number that is forwarded to the 911 *Public Safety Answering Point (PSAP)* is the *Billing Telephone Number (BTN)* used for billing the Type 1 connection and the street address used is that of the MTSO or Point Of Termination. There are methods available to identify the originating cell site to the PSAP operator but there is no way to automatically identify the physical location of the mobile subscriber at the time the 911 call is placed. Essentially, this is why the agreement of the local public safety agency must be secured before 911 calls are forwarded to the LEC network.

Operator Services Traffic

Due to some unresolved business and technical issues, access to a LEC operator (0-) or the ability to place a call on either the LEC network or via an IC using a calling card is presently restricted to a Type 1 connection.

IC operators may be accessed through either a Type 1 or a Type 2A connection. Using the Type 1 connection, the IC operator of the presubscribed carrier is reached by dialing 00, or 10XXX+0 for all other ICs.

ROAMING WITH A CELLULAR TELEPHONE

A cellular subscriber obtains service from a CMC in a given location and the information that is electronically stored in the mobile unit (MIN, ESN, and SI) identifies that subscriber's "home" system. When the cellular subscriber travels to other areas, the subscriber is said to be "roaming." Service can then be provided by the "Visited" system.

In order to properly bill for calls made in other systems, agreements are made between CMCs to exchange billing data and accept some liability for calls made by the roaming subscriber. A third party, called a "Clearinghouse," is currently used to validate certain data and perform the necessary data exchange.

Mobile-To-Land Calls—Roaming

The process is similar to calls in the home system except that some verification occurs after the presence of the roaming mobile unit is detected. Because it receives the MIN, ESN, and System Identification, the MTSO is aware that the unit making the call is in the roaming condition. Likewise, the mobile unit displays a "roaming" indication because of the system identification information it has received from the base station over the control channel. Usually, the first call from a roaming mobile unit is permitted to continue without interruption although the MTSO initiates an inquiry to the clearinghouse via a separate data link. The purpose of the inquiry is to determine if the information received is from a valid mobile unit and a credit check is also made at this time. If the response to the inquiry indicates the mobile unit may be fraudulent or a potential credit problem exists, the MTSO routes any future calls from the mobile unit to an announcement. The announcement advises the mobile unit to contact the business office of the "roaming" CMC for further instructions. While these inquiries are currently not done on a real-time basis, the capability to eliminate this limitation is almost developed.

Land-To-Mobile Calls—Roaming

Calling a mobile unit that has "roamed" to another area can be done in one of two ways at the present time. The first method involves the use of a "roamer port" while the other provides the appearance of automatically locating the roaming unit.

Using a roamer port, the original method, is still used in some instances. Using the roamer port is cumbersome because the landline caller must know where the mobile unit is roaming, which system (wireline or non-wireline) is being used, and the roamer port number of the correct system. Assuming the landline caller has all of that information, a call is placed to the roamer port telephone number. Typically, roamer port number ends in the digits 7626 so that it spells ROAM. An example might be 201-740-7626 (or 201-740-ROAM). This call is often a toll call so the landline user will be charged for any applicable toll charges. Upon reaching the roamer port, a second dial tone is provided by the MTSO. The landline caller must then input the actual telephone number of the mobile unit. The MTSO then pages the mobile unit on the control channel and the call is set up like any other land-to-mobile call.

Because the original method of locating a roamer requires a great deal of sophistication and effort by the landline customer, General Telephone and Electronics (GTE) developed a concept called *Follow Me Roaming* that eliminates any additional effort by the landline subscriber. Any applicable toll charges are paid by the mobile subscriber. The mobile subscriber activates the *Follow Me Roaming* feature upon reaching the roaming area by pressing "*18" on the mobile unit. At this point, the Visited MTSO receives the signal and initiates an inquiry to the clearinghouse to verify the mobile unit. In addition to the normal inquiry, the Visited MTSO provides the clearinghouse a *Temporary Local Directory Number (TLDN),* which is then forwarded to the Home MTSO. Similar to the Call Forwarding feature in the landline network, the TLDN is used by the Home MTSO to route calls through the PSTN. It is valid until the time period established by the Visited MTSO expires (usually a maximum of one day) or until the mobile subscriber cancels the *Follow Me Roaming* feature by dialing "*19." A landline caller attempting to reach the mobile unit simply dials the mobile unit's normal telephone number. The Home MTSO, aware that the mobile unit is roaming in the Visited System, routes the call to the Visited MTSO using the TLDN it has received from the clearinghouse. Upon receiving the call, the Visited MTSO associates the TLDN with the roaming unit and initiates a page over the control channel using the mobile unit's normal telephone number. Call set-up continues from that point as with any other call.

6 PAGING SYSTEMS

SERVICE EVOLUTION

An irritated hospital patient, World War II, and the man who inspired the two-way wrist-radio for Dick Tracy. These are all related to the start of the paging services that have become so popular in the United States, and now worldwide.

The radio paging unit was invented by a very talented radio pioneer named Al Gross. According to Gross, the concept of paging service originated in 1939 when a man named Charles Neergaard was a patient in a hospital for over two months. Neergaard was annoyed by the incessant messages that were broadcast over the hospital loudspeaker system. He conceived the idea of silent paging using radio transmission as an alternative but, although there was interest in his idea, he could not get anyone to turn his idea into reality.

Part of Neergaard's difficulty was the technological limitations in 1939. However, during World War II, great technological advancements were made, including a secret, and very compact, radio that was successfully used by secret agents for clandestine air-ground communication. This hand-held transceiver, which was hardly larger than current cellular phones, was designed by the ingenious Al Gross. During the war, widespread military use of radio technology led to considerable press coverage of these devices. After the war, Gross developed a hand-held Citizen's Band radio, which was so small for its time that the author of the *Dick Tracy* comic strip was inspired to include a two-way wrist-radio for his character, and started using the idea in a segment published in 1948.

In 1950, with help from Neergaard and others, Gross designed and developed a small, one-way radio paging receiver, and a companion 100 watt transmitter, for use in hospitals. In 1952, Gross received the first certification from the *Federal Communications Commission (FCC)* for one-way paging systems for use in hospitals.

Gross' device was a great advancement because it provided a means of selective radio signaling. It operated in the 460-470 MHz band, which made it suitable for use inside buildings.

Because many people really needed only one-way communication, and the available radio spectrum is limited, the paging industry began to expand rapidly in the 1950s as a means of reducing the demand for two-way communication. Using an individual channel for one-way paging service allowed the same amount of spectrum to provide service for more people than if it were used for two-way communication.

As a result of an FCC decision in 1949, additional radio spectrum was allocated for one-way paging service. Competition was introduced by allowing entities besides *Local Exchange Carriers (LECs)* to also provide these services. These entities, originally called *Miscellaneous Common Carriers (MCCs),* are now known as *Radio Common Carriers (RCCs).* Aircall, Incorporated, which began service in New York in 1952, was the first RCC to be licensed by the FCC for paging service. Today, the RCCs provide the vast majority of paging services in the United States.

All the early radio paging systems required the use of an operator to complete a paging call. Automatic dial arrangements began to appear in the late 1950s, beginning with an experimental system by Bell of Pennsylvania in its Allentown location in 1957. Now virtually all paging systems, whether provided by RCCs or the LECs, use automated systems.

Types Of Paging Services

Paging services currently available to subscribers include tone-only, tone-voice, alphanumeric, and visual display operation. Each service represents an evolution in technology and each has an appeal to its segment of the paging market.

Tone-only service allows more users to be served per paging channel and permits the use of smaller, simpler, and cheaper paging receivers. Tone-only is simply what the name suggests—a tone is emitted which alerts the user that someone is trying to reach him/her. It is also possible to have two distinctive tones so that each implies a different meaning to the user.

Improvements were made to the receivers to enable tone-voice service. Tone-voice systems continue to be a very popular service with paging subscribers. It allows the user to receive and assess very specific pages.

Although a limited form of display paging was possible with analog paging systems, it was the introduction of digital systems that made alphanumeric and display service practicable. These devices have progressed from simply displaying numeric information to having the ability to transmit several pages of text that include both alpha and numeric data.

RADIO SPECTRUM REQUIREMENTS

Paging service is provided using several frequency bands authorized by the FCC. Early systems were limited to the 35 and 43 MHz bands, and a number of systems using these frequencies are still in use. Because of the "skipping" characteristic of these frequencies over long distances, rather stringent rules are used to ensure adequate separation between systems. These rules permitted channels to be reused at distances up to a few hundred miles, but they could not be reused at distances from about 1000 to 3000 miles. As a result, the availability of paging service was reduced in some areas. In addition, while the 35 and 43 MHz systems provide good coverage in open areas, they are not well suited for penetrating buildings to reach users.

Later, the FCC authorized paging service using frequencies in the 150 and 450 MHz bands. These same frequencies are also used for two-way mobile service. The FCC eventually allowed the RCCs to utilize these frequencies for either service, or to share them, as required for their business needs. The Bell System began introducing paging systems using the 150 MHz band in 1960.

Recognizing the need for a nationwide paging service, the FCC authorized frequencies in the 931 MHz band for this purpose. A proceeding concluded in 1985 authorized three nationwide carriers, each selected by lottery, to provide this service. Local RCCs can affiliate themselves with these nationwide carriers. Many nationwide carriers use satellite facilities as a link to all cities in order to reach the customer being paged.

BASIC ARCHITECTURE

A typical paging system employs one or more transmitters to cover an area defined by the FCC as the *Reliable Service Area (RSA)*. It differs from the *Rural Service Area (RSA)* terminology used for cellular service because it defines an area of reliable radio transmission instead of a geographical area determined by county boundaries. Depending upon a number of factors, the *effective radiated power (ERP)* of the transmitter can be as high as 500 watts. Figure 6.1 illustrates a typical paging system.

Typical Paging System
FIGURE 6.1

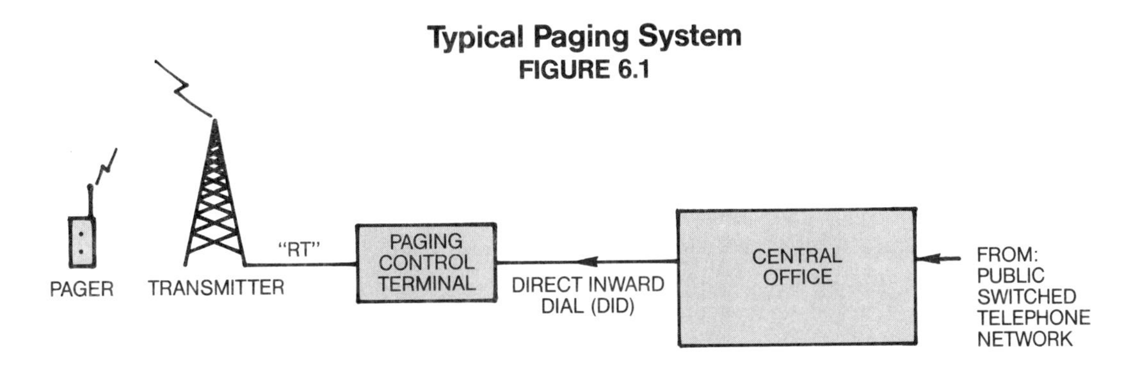

Paging Control Terminal

A major component of a paging system is the control terminal which receives the digits that have been dialed to reach the pager from the LEC network. The control terminal also verifies that the number received is a valid number on the system, routes calls to non-valid numbers to a recorded announcement, sequences the calls for transmission, and controls the radio transmitter(s). It also provides the necessary supervisory signals to the LEC switch and translates the telephone number received from the LEC network to the appropriate format for transmission over the paging system.

Paging Formats

A specific format did not really exist for early paging receivers because they were simply radios tuned to a particular frequency. Due to the limited number of channels available, the users of these devices received many messages that were intended for other subscribers on the system. To overcome these limitations, selective signaling formats have been developed to increase the efficiency of the paging system while simultaneously providing privacy to the paging user.

Two-Tone Signaling

The earliest of selective signaling formats, two-tone signaling, used two discrete tones that were transmitted sequentially to alert a paging subscriber. A variety of manufacturers developed their own versions of two-tone signaling formats. While the concept was the same, the formats basically differed in the duration of the tones and the frequencies that were used. This system was relatively slow, with some models requiring almost five seconds of airtime to alert the selected pager.

Three-Tone Signaling

Used extensively by the Bell System for its Bellboy™ service, the control terminal would convert the last four digits of the paging unit's telephone number into three of 32 possible audio tones. Initial versions of the paging receiver were equipped with special reed selectors that responded only to these three tones, thus providing the desired selectivity and privacy. Later versions used active, electronic filters in place of the reed selectors. To improve efficiency over the two-tone method, the three tones were transmitted simultaneously instead of sequentially. However, some of this speed advantage was reduced because each tone was transmitted three different times in order to increase the probability of being received. While each transmission consumed about 0.8 seconds, due to the repeat transmissions, the total time required was almost three seconds.

Five-Tone Signaling

Still used in analog systems, this format converts the received telephone number to a five-tone format that uses five out of a possible 12 tones. Ten of these 12 tones signify the digits 0 through 9 while an eleventh tone, called "R," is used to indicate that a digit is to be repeated. The twelfth tone, labeled "X," is sometimes referred to as a sixth tone. It permits the pager to emit a different beeping pattern for tone-only paging. The five-tone system reduces airtime to about 0.22 seconds per page.

Golay Sequential Coding (GSC)

Digital paging systems were developed in the 1970s. Motorola introduced a format called *Golay Sequential Coding (GSC),* in honor of its developer. Used widely today, and often referred to as simply Golay, this format permits alphanumeric characters to be transmitted for those customers wanting an alphanumeric display. For tone-only or voice and tone service, digital paging using the Golay format is faster than its analog predecessors, making the paging system more efficient. For example, with tone-only paging, Golay completes a page in about 230 milliseconds, or roughly four pages per second. This system includes error correction to reduce the occurrence of false or missed pages. It is also in use outside of North America.

Post Office Code Standard Advisory Group (POCSAG)

Developed by the British Post Office and officially called the *Post Office Code Standard Advisory Group (POCSAG),* this format is now an internationally accepted standard for signifying a pager's electronic address. Like the Golay format, POCSAG offers error correction. Because any manufacturer can design equipment based on this standard, the use of POCSAG allows small manufacturers to produce competitive equipment. In an effort to avoid duplication of pager address codes, Telocator, a wireless industry trade association that represents paging and other wireless carriers, administers the assignment of the POCSAG codes for paging devices. However, it is not mandatory that Telocator is notified each time a code is encoded into a pager. As with its Golay competitor, an alphanumeric display can be provided using the POCSAG format.

Paging Receiver Units

Paging receiver units have become increasingly smaller. Al Gross' first tone-only pager was about the size of a king-size pack of cigarettes and weighed 12 ounces. Because of its vacuum-tube technology, its non-rechargeable batteries had to be changed often. Tone-only pagers today are smaller than a small cigarette lighter and weigh only a few ounces. A small battery can power the pager for several months. Even the sophisticated alphanumeric display pagers are much smaller than the original tone-only receiver. There is even a true wrist watch model with a digital display now available.

The signaling frequency of all early paging receivers was fixed to a specific frequency band. This limitation is still present in many modern paging receivers, but models are available that permit the user to select a frequency band. This feature is particularly useful for subscribers who travel and want service in a number of different geographical locations. In an effort to conserve battery power, some units have a "sleep" feature. Pagers with this feature power-down most of the circuitry to conserve energy. When a message is broadcast for a pager with this feature, a preamble in the message alerts the pager and restores full power so it can receive the message.

Radio Transmitters

Paging systems utilize *frequency modulated (FM)* transmitters. A system may employ a single transmitter, as illustrated in Figure 6.1, or have multiple transmitters in order to provide adequate coverage in its service area. When multiple transmitters are used, the signal is transmitted almost simultaneously by all of the transmitters, giving a *simulcast* system. With such a system, special time-delay equalization is used to ensure that the paging signal leaves all transmitters at the same instant, to prevent interference to a receiver that is served by two transmitters. An alternative is to send the page over each transmitter in sequence, at the cost of added airtime.

The transmitter (or transmitters) may be remotely located from the paging control terminal and is operated by an inband signal supplied by the control terminal. Older systems required a *direct current (dc)* path for this transmitter-keying operation.

The output power of a paging transmitter can vary, depending on the desired geographic coverage and antenna height. A limit of 500 watts *effective radiated power (ERP)* is imposed by the FCC for most frequencies.

The FCC rules for FM broadcast stations permit the use of a subcarrier signal for paging. Transmitted along with the FM programming, the signal is detectable by special pagers operating in the 88-108 MHz FM band.

Some paging systems share base stations and radio channels with two-way mobile systems. To prevent interference to the mobile voice users, the mobile receivers are equipped with a tone-squelch feature that disables the receiver unless a mobile conversation (with the tone signal) is in progress.

System Capacity

The capacity of a paging system is highly variable. Depending upon spectrum availability and equipment configurations, a typical paging system can support 500 to 50,000 subscribers.

NUMBERING REQUIREMENTS FOR PAGING SYSTEMS

In a dial paging system, each paging unit is identified by a unique address that conforms to the *North American Numbering Plan (NANP)*. The paging carrier providing the service obtains blocks of numbers from the LEC and assigns these numbers to its paging subscribers. These numbers may be part of an NXX code that the LEC uses to serve its own subscribers, meaning the NXX code is shared with other users. However, if the paging carrier's requirements are large, the use of an NXX code dedicated to the paging carrier is possible.

If a shared code is used, the blocks of numbers should be as sequential as possible. In earlier systems, such as the Bellboy™ system, there were additional constraints in that the numbers could only be assigned from certain groups because of the technique used to translate the telephone number to a three-tone code. Such restrictions are not common with more modern systems.

Similar to the procedures outlined for cellular service in Chapter 5, the location of the NXX code for billing purposes is either the LEC end office or the *Point Of Termination (POT)* at the paging carrier's location. Unless a Type 2 connection is used, the *Vertical and Horizontal (V&H)* coordinates used for billing purposes will be those associated with the LEC end office.

INTERCONNECTION TYPES

Paging systems use *Direct Inward Dialing (DID)* circuits almost exclusively for connecting their systems to the *Public Switched Telephone Network (PSTN).* These trunks permit the LEC end office to outpulse from four to seven digits of the telephone number that was dialed to reach the paging subscriber.

As a result of the FCC's declaratory ruling in 1987, RCCs providing paging service may obtain the same interconnections as used by the cellular carriers, i.e., Type 1, 2A, or 2B. As a practical matter, most paging systems are designed to operate using DID trunks for the incoming (land-to-pager) calls.

Paging carriers may also obtain private line circuits from the LECs to connect the paging control terminal to remotely located radio transmitters. These private line links are usually voice-grade analog circuits, but digital private lines at the DS1 (1.544 Mbps) rate are also possible.

INTERFACE REQUIREMENTS

Two-wire analog interfaces are presently the most common form of interface obtained by carriers offering paging service. However, digital interfaces are beginning to become more prevalent because paging terminal equipment is now available to accept a digital signal. Also, the LECs have substantially reduced their prices for digital facilities in recent years.

CALL PROCESSING

Radio paging is, by definition, a one-way service. As such, only the processing of land-to-paging calls is a concern. While earlier systems were manual operations requiring the intervention of operators to complete the call to the paging subscriber, practically all systems in the United States use automatic dial equipment. The sequence described below assumes an automatic system.

To reach a paging subscriber, the telephone number assigned to the paging unit is dialed. The call is routed through the PSTN to the LEC office serving the paging carrier. Depending on the interconnection type that is employed by the paging carrier, this LEC office may be an end office (for dial line, DID, or Type 1 connections) or a tandem office (for Type 2 connections). Regardless, the LEC office outpulses from three to seven digits of the dialed number to the paging carrier.

Upon receipt of the outpulsed digits, the paging control terminal determines whether the digits represent a valid paging subscriber or an unassigned number within the number group obtained from the LEC. If the latter is the case, the call is routed to an announcement to inform the caller that he has reached a non-working number. If the call is to a valid paging subscriber, the control terminal returns answer supervision to the LEC office. In addition, it responds with a distinctive "tinkle" tone, and sometimes a voice announcement, to advise the caller that the call has been received and will be transmitted. The call is then placed in a queue by the controller to translate the telephone number into the appropriate paging format and to await transmission. In response to inband signals generated by the paging control terminal, the transmitter (or transmitters) broadcast the signal. To further ensure that the paging signal is received, it can be transmitted two times in rapid succession. This is called *repeat page* and is generally offered as an option, at a modest cost, to the paging subscriber.

When the transmitted signal reaches the paged receiver, the information is decoded causing the unit to emit a "beep," or in some cases, initiate a vibrating action, to alert the user to check the unit or respond with a prearranged action.

7 AIR-GROUND SYSTEMS

SERVICE EVOLUTION

There are two types of air-ground systems in the United States. One is for general aviation aircraft, which includes private or corporate airplanes. The other is used for passengers on commercial airliners. Both provide the capability for airborne users to complete calls to landline users via connections to the *Public Switched Telephone Network (PSTN).* Presently though, only the general aviation systems have the capability to complete land-to-air calls.

Air-ground service for general aviation aircraft began on an experimental basis in 1957 when Illinois Bell began operating a *base station* in Chicago and, simultaneously, Michigan Bell opened a station near Detroit. Service was expanded to other cities and in 1970, the *Federal Communications Commission (FCC)* ordered the service to be provided on a *common carrier* basis. Consequently, air-ground service is provided by *local exchange carriers (LECs)* and *radio common carriers (RCCs).* Presently, there are 84 air-ground stations in operation in the United States. Approximately 60% of these air-ground stations are operated by RCCs. One RCC, Mtel, operates over 25 stations.

Originally, all of the general aviation air-ground systems were completely manual. Today, over 90% of the stations are fully automated, offering a service known as the *Air/Ground Radiotelephone Automated Service (AGRAS).*

Air-ground service for commercial aviation was introduced by Airfone, on an experimental basis, in 1984. Airfone was subsequently acquired by *General Telephone & Electronics (GTE)* and is now known as GTE-Airfone. As part of its on-going effort to introduce competition, the FCC issued an order that made it possible for other entities to provide commercial air-ground service. Presently, six other carriers have applied for licenses to compete with the existing GTE-Airfone service.

Interestingly enough, with the exception of Canada and the Virgin Islands, the United States is the only country in which air-ground service is available.

RADIO SPECTRUM REQUIREMENTS

Air-ground service for general aviation uses the same *Ultra High Frequency (UHF)* frequency band as is allocated for common carrier two-way mobile service. Operating in the 450-MHz band, each channel provides 25 kHz of *bandwidth,* which yields a total of 12 voice channels and 1 signaling channel. A maximum of 4 voice channels may be used at any one station, but many stations require only one voice channel.

Miami is the only city currently using all 4 voice channels while some large cities, like New York, have 3 active channels available. In order to avoid *co-channel interference* at altitudes as high as 39,000 feet, stations having the same channel assignments are separated by FCC rules by distances of approximately 550 miles or more.

A total of 4 MHz of spectrum within the 800-MHz band was allocated by the FCC in 1990 for commercial air-ground service. This spectrum was divided into 10 channel blocks, each occupying 200 kHz of spectrum. Although debate continues on some of the details of the plan, the current proposal anticipates each block will contain 31 communication channels, 4 pilot channels that are used for control purposes, and 1 guardband channel that is used to separate the communication and pilot channel frequencies. The voice channels are used for communications. The FCC has assigned one channel block to specific geographical areas throughout the United States. These sites are located in 40 of the 50 states with most of the omitted states being in the northeastern part of the country. Because the stations are designed to serve relatively large areas, there is no need to have a station in every state. As with the general aviation air-ground service, areas using the same frequencies are separated by a minimum of 550 miles in order to avoid co-channel interference.

The communication and guardband channels within any particular channel block will be shared by all of the air-ground carriers. As presently proposed, each licensed carrier would be assigned its own control channel. In the event insufficient control channels are available, the FCC proposes to assign a specific communication channel to a carrier for use as a control channel.

BASIC ARCHITECTURE

Although they serve different segments of the aviation industry, the system architectures of both types of air-ground systems are very similar. Each consists of a number of ground stations containing the system controller and transceiver(s), and airborne transceivers. Each system is also interconnected with the PSTN. Figure 7.1 illustrates a typical air-ground system.

Each system was also designed to cover a geographical area that ranges from 100 to 300 miles. Because general aviation aircraft often operate at lower altitudes than commercial aircraft, coverage is not totally ubiquitous. Neither system possesses a hand-off capability because they were not designed for lengthy conversations. However, conversations that last 20-30 minutes are possible with each system.

Base station (or ground) transmitters may transmit at a maximum *effective radiated power (ERP)* level of 100 watts while airborne units are limited to no more than 25 watts ERP for general aviation and 30 watts ERP for commercial aircraft.

Typical Air/Ground System
FIGURE 7.1

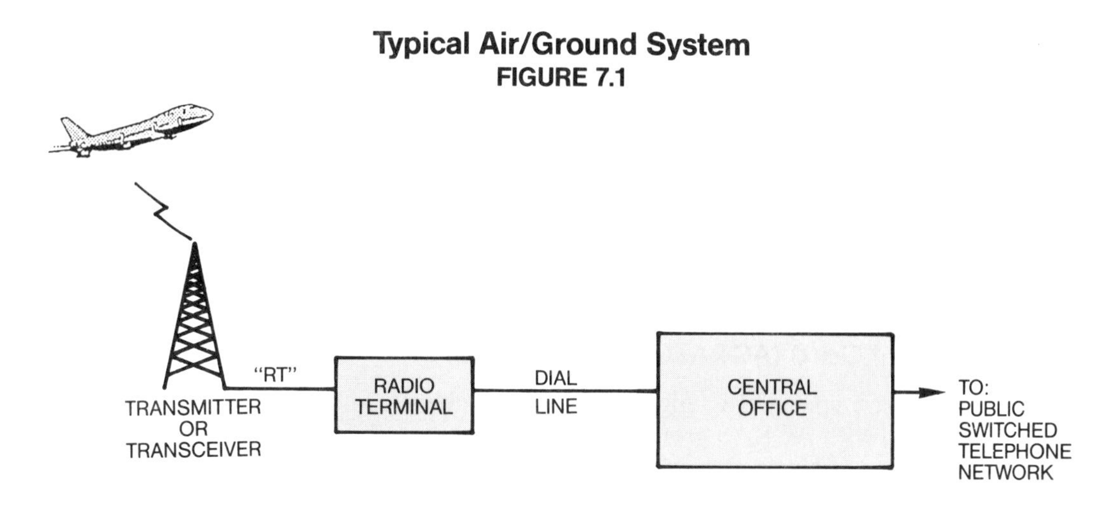

The system controller for an air-ground system performs the same functions as a system controller in a two-way mobile system. In fact, some of the early air-ground systems used the same equipment for this purpose as was used for the two-way mobile applications. The system controller determines if the user is authorized to use the system, allocates a radio channel, and records the call for billing purposes.

NUMBERING REQUIREMENTS FOR AIR-GROUND SYSTEMS

Currently, neither type of air-ground system requires the assignment of telephone numbers to individual aircraft units. This is because traditional telephone numbers are not used to identify aircraft units. Instead, special numbers are used for identifying and billing calls involving aircraft units. For general aviation aircraft, these are *QM* numbers and a special type of credit card number called an *Air/Ground Radiotelephone Automated Service Credit Card Number (AGRASCCN).* For commercial aircraft, each unit is identified by an unique number issued by the carrier (which currently is GTE-Airfone).

QM Numbers

Used only for general aviation aircraft, a QM number must be assigned before the FCC will license a particular unit. The letters "QM" evolved from early two-way mobile systems, which assigned telephone numbers for the mobile units using letters, JL, QD, QM, etc. that could not be confused with central office names. A QM number is an 8-digit number that identifies the airborne unit. It is assigned to the aircraft instead of the individual who owns the aircraft. The QM number is encoded into the set and, particularly for manual systems, allows the unit to be called selectively by landline users. The first 5 digits of the QM number provide the unique identity of the set while the last 3 digits identify the *revenue accounting office (RAO)* of the *local exchange carrier (LEC)* that performs the billing function for these calls.

Bell Communications Research (Bellcore) allocates blocks of QM numbers to the *Mid-America Computer Corporation (M.A.C.C.)* and the LECs. M.A.C.C. and the LECs assign the QM numbers upon request by aircraft owners. M.A.C.C. is a firm that specializes in billing for air-ground service and actually assigns the majority of QM numbers.

Bellcore publishes a directory of assigned QM numbers that is based on data received from the M.A.C.C. and the LECs.

AGRAS Credit Card (AGRASCCN) Numbers

AGRASCCNs are special 18-digit credit card numbers, issued by the M.A.C.C., used with AGRAS ground stations. These numbers are also encoded into the airborne sets and are used for billing air-ground and ground-air calls.

INTERCONNECTION TYPES

Present versions of air-ground systems primarily use dial line connections for interconnection with the PSTN. Because the aircraft have unique identification numbers, and the service is primarily unidirectional (air-ground), other connection types are not necessary. However, because these carriers are classified as common carriers by the FCC, they are entitled to the same types of interconnection as cellular carriers (Type 1 and Type 2).

If the receiver or transmitter antennas are remotely located from the system controller, the air-ground carrier may use private line, voicegrade *Radio Telephone (RT)* circuits to connect the system controller with these remote locations.

INTERFACE REQUIREMENTS

Because these systems require only small quantities of dial line connections, two-wire analog interfaces are almost exclusively used by carriers providing this service.

CALL PROCESSING

Due to differences in the manner in which calls are billed, and the fact that commercial air-ground service is presently limited to one-way (air-ground) calls only, there are differences in the call processing methods used by each system. One common feature with both systems is that modern airborne equipment automatically selects the ground station delivering the strongest signal. Each system also is designed to attempt to select the ground station in front of the aircraft in order to provide maximum call duration.

Air-Ground Calls With Automated Systems—General Aviation

With automated systems, calls are processed much like a landline call. If the aircraft is in range of a ground station (usually about 100 miles), a lamp will indicate a channel is available for service. The caller dials the desired digits, which are transmitted to the ground station's receiver. The system controller verifies the user data (AGRAS number or QM number), seizes a dial line connection to the PSTN, and outpulses the desired digits. The call is then processed through the PSTN as with any other call. The system controller records the call details and a bill is ultimately rendered to the user which includes the airtime as well as any PSTN charges. The airborne user may have the PSTN charges billed to a credit card issued by a LEC or an *interexchange carrier (IC)* by prefixing the dialed digits with a "0" instead of the digit "1." The airtime charges are still billed to the AGRASCCN or QM number.

If all of the ground station channels are busy, the airborne set usually emits an alternating tone, sometimes described as a "French siren." When a channel becomes available, the airborne set will ring indicating the call can now be placed.

Air-Ground Calls With Manual Systems—General Aviation

If the nearest ground station is a manual system, the user will hear a high-pitched tone indicating a channel is available. If no tone is heard, service is not available at all. If all channels are in use, voices will often be heard unless a set equipped with a privacy feature is being used. Once a vacant channel is selected, an operator will request billing information (AGRASCNN, QM number, or both), and the telephone number being called. The operator then seizes a dial line connection, and dials the desired number. The call is then processed like any other call.

Air-Ground Calls—Commercial Aircraft

Calls from commercial aircraft equipped with air-ground units require the use of a standard credit card (Visa, Mastercard, etc.) for billing purposes. To initiate a call, the user places the credit card in a slot on the unit. The airborne unit automatically selects a channel, if available, from a ground station ahead of the aircraft. Because of the high altitudes used by commercial aircraft, the ground station may be up to 300 miles in front of the airplane. The credit card information is transmitted by the airborne unit to the ground station. After verifying the credit information, the system controller sends a signal to the airborne unit indicating service is available. The user then dials the digits of the party being called, which are transmitted by the airborne unit to the ground station. The system controller then seizes a dial line and outpulses those digits to the PSTN so the call may be completed. All charges, including airtime and any PSTN charges, are billed to the user's credit card.

Ground-Air Calls With Automated Systems—General Aviation

It is possible for a landline user to call a general aviation aircraft in flight. The process is similar to "roaming" calls to two-way mobile sets because the person making the call must know the approximate location of the aircraft, the telephone number of the ground station in that area, and the aircraft's AGRAS or QM number. The landline caller dials the number of the ground station closest to the aircraft, and receives a second dial tone. Upon receipt of this second dial tone, the user then dials the QM number (or for AGRAS-equipped stations the first 10 digits of the AGRASCCN). The system controller then selects a vacant channel and broadcasts the QM or AGRASCCN number on the signaling channel. If the aircraft is within range, the airborne unit responds and is directed to the proper channel by the system controller. The airborne set then rings to alert the airborne subscriber to an incoming call.

8 PERSONAL COMMUNICATION SERVICES CONCEPTS

Personal Communication Service (PCS) is a term that has recently attracted a great deal of attention. While a precise definition of PCS does not yet exist, many believe it will provide customers with communication options that will allow a great deal of personal freedom and mobility. PCS can be thought of as a wide array of telecommunication services that includes the existing *Public Switched Telephone Network (PSTN)*, the wireless services already discussed in the previous chapters (two-way mobile, cellular, paging, and air-ground), as well as new possibilities. Two of these new possibilities are concepts labeled *CT2 (Cordless Telephone—2nd Generation)* and *PCN (Personal Communications Network)*.

CT2 SERVICE EVOLUTION

As the name CT2 suggests, there is a predecessor called *CT1 (Cordless Telephone—1st Generation)*. The first generation constitutes those cordless telephones used in millions of households and small businesses. These phones have a base station (an especially apt description in this case since this is where the handset rests when not in use) and a mobile station (the handset). In the United States, these devices operate mainly in the 47-49 MHz band. They are sold to individuals and are not part of a wireless service provided by a carrier. Because they operate at very low power levels under Part 15 of the FCC rules, and each individual in essence "owns" the CT1 system, there is no need for each user to obtain a license to operate these devices.

These cordless telephones are designed to provide service primarily within about 50 feet from the base station location. Very early models provided no channel selection or speech encryption of any type. Thus, they provided little privacy and were subject to fraudulent use because others could access a base station using a handset that operated on the same frequency. Later models have improvements to reduce these problems.

Building on the limited freedom afforded by the first generation of cordless telephones, a concept of using a cordless phone in public areas was initiated in the United Kingdom. This concept, called CT2, provides limited mobility by allowing a user to access the PSTN through public base station locations provided by a wireless carrier.

In 1989, the United Kingdom's *Department of Trade and Industry (DTI)* issued four licenses for CT2 service. These licenses were held by consortia named Mercury Callpoint, Phonepoint, Ferranti Creditphone, and Hutchinson Telecommunications. Although marketed under different names, each system was initially conceived to be a pay-station alternative configured to provide one-way access to the PSTN. This was done by constructing base stations in public places that were identified by the logo of the provider (Phonepoint, Rabbit, etc.). Special cordless telephones could obtain service from these sites as long as the user was within about 600 feet of the radio location.

It was envisioned that the initial CT2 configuration would eventually have limited two-way capabilities. However, the one-way service met with very little commercial success. A major reason for this failure was that the service was inaugurated before any standard air interface was established. Therefore, the sets were incompatible between systems and the users were limited to obtaining service only from sites provided by a particular carrier. While a standard was eventually developed, this initial deficiency, coupled with the small number of sites constructed, has stymied introduction of the service. As of January 1992, three companies that began service have suspended operations. The fourth, Hutchinson, has yet to begin providing service.

In the United States, market research by several firms has indicated a strong desire for CT2-like service that would provide a degree of mobility at a price considerably less than that of present cellular service. A number of trials are underway to test the concept from both a technical and market perspective. A standard for the air interface is being discussed and it is one of the goals the FCC is addressing in an ongoing PCS proceeding.

It is quite possible, and indeed very probable, that CT2 will progress beyond the original concept of a one-way, pay-phone substitute service. It will likely become a two-way service that will permit calls to be made or received within the subscriber's home, neighborhood, or selected public places.

CT2 RADIO SPECTRUM REQUIREMENTS

Currently, no decision has been made by the FCC regarding the radio spectrum requirements for CT2 service in the United States. In 1990, a request for 2 MHz of bandwidth in the 900 MHz band was made by Cellular 21. This request has been incorporated into the FCC's Docket 90-314 proceeding on PCS.

Prior to the cessation of operation in the United Kingdom, each carrier was allocated 4 MHz of bandwidth within the 800 MHz band. Using a *frequency division multiple access (FDMA)* modulation scheme, this yielded 40 voice channels per operator. With this technology and bandwidth, each system was expected to support approximately 12,000 subscribers per square mile.

Another standard being discussed in Europe for CT2 applications is called the *Digital European Cordless Telecommunications (DECT).* This technique uses *time division multiple access (TDMA)* instead of FDMA and proposes operation in the 1.88-1.90 GHz band. This would potentially support 20,000 to 200,000 users per square mile.

Experimental systems in the United States are utilizing a number of different frequency bands while testing various modulation techniques.

BASIC CT2 SYSTEM ARCHITECTURE

CT2 is a relatively simple architecture because it is an access technology and not a self-contained network. In this sense, it resembles air-ground service which uses the PSTN, rather than the cellular network which can operate as a self-contained network. CT2 is designed to allow access to the PSTN, but it is possible that future CT2 arrangements could access other networks instead of the present PSTN. Architecturally, a CT2 system has a number of small radio ports, a central database, and mobile sets used by the subscribers. A typical system is illustrated in Figure 8.1.

The radio ports are located at various points to serve public places, such as airports, shopping malls, and other pedestrian locations. It is also possible that some sites may serve private residences or businesses as well. These radio ports are designed to serve small geographic areas of up to 1000 feet in radius. These locations are like small islands rather than contiguous cells, and there is no handoff capability between sites. The radio port will normally contain a radio controller function, but this function could be remotely located from the radio site. The site will provide access to a number of radio channels, or ports, whose number depends on the expected traffic load. The radio channels are allocated by the controller, which also validates the user, connects the call to the PSTN, and records billing details.

A central database periodically collects information from all of the radio port sites so that usage for subscribers is aggregated and billed. The central database also revises subscriber information stored at the radio sites, providing the most current information for use by the sites to validate subscribers attempting to obtain service.

Mobile sets used by CT2 subscribers are very low powered devices that transmit at a maximum power of 10 milliwatts, or 1/100th of a watt. It is anticipated that these sets will be relatively inexpensive.

CT2 NUMBERING REQUIREMENTS

Although the CT2 equipment in the United Kingdom did not require the use of traditional telephone numbers, it is expected that CT2 service in the United States will ultimately have two-way capability. For this reason,

Cordless Telephone—2nd Generation (CT2) Concept
FIGURE 8.1

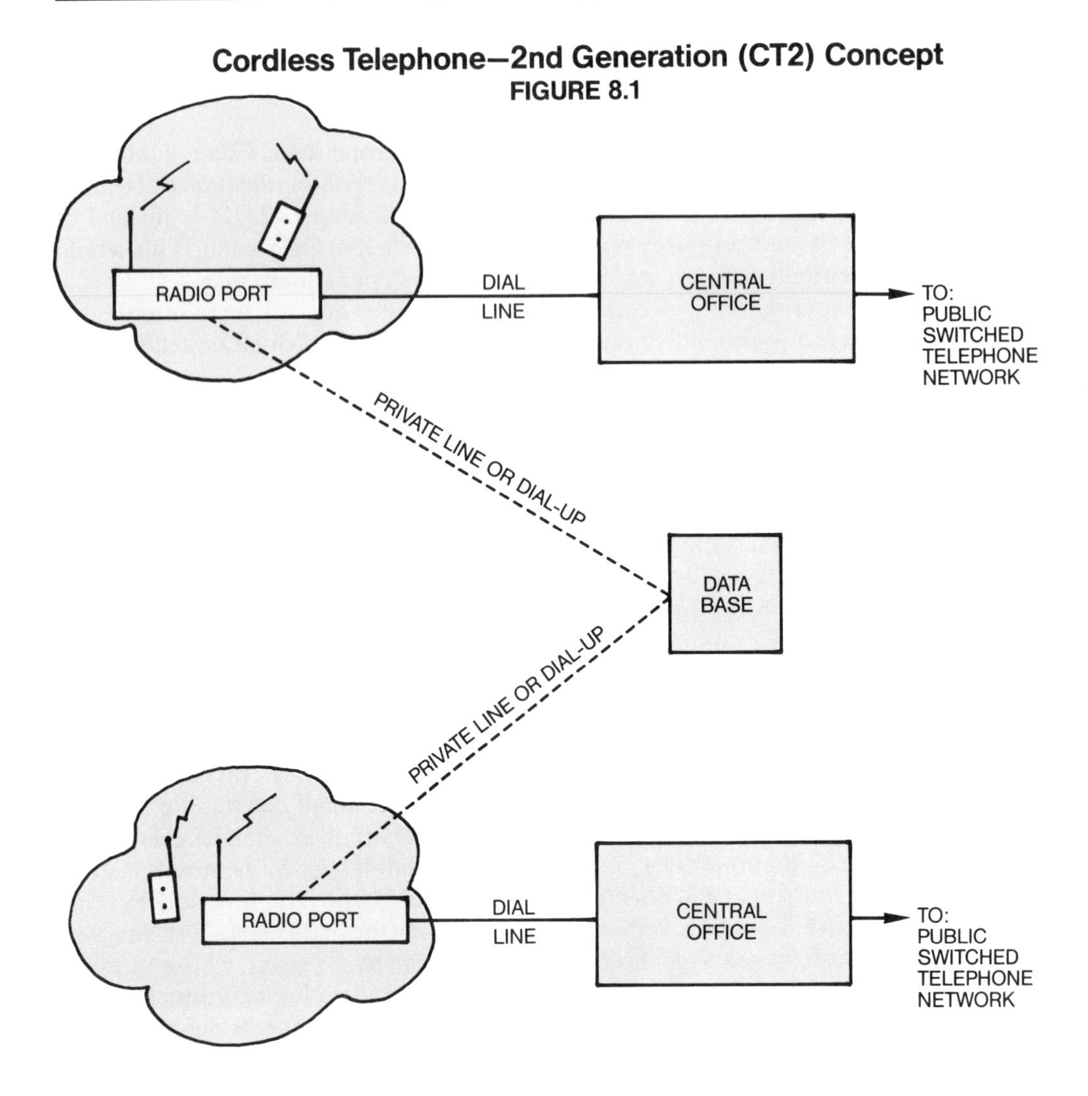

a number conforming to the *North American Numbering Plan (NANP)* will probably be associated with the CT2 set. This number may serve a function similar to that of the *Mobile Identification Number (MIN)* for two-way mobile and cellular systems. It is also likely that the sets will also contain an *Electronic Serial Number (ESN)* as well as a *System Identification (SID)* number. Some of this functionality may be provided through a "smart card" inserted by the user, which would enable a set to be utilized by several users with different telephone numbers.

NXX codes used for CT2 service may be shared with other services or dedicated to the exclusive use of the wireless carrier providing the CT2 service. Whether shared or dedicated codes are used will depend on the total numbering requirements of the carrier.

INTERCONNECTION TYPES USED FOR CT2 SERVICE

CT2 systems in Europe, as well as the experimental systems in the United States, have used dial line connections from a *local exchange carrier (LEC),* like PBX trunks. This connection from the radio port location provides suitable access to the PSTN. Whether other connection types will be utilized will depend on how CT2 is eventually classified by regulatory agencies as well as any changes in the technology used to provide the service.

If the radio controller is not collocated with the radio port, a private line circuit, like a *Radio Telephone (RT)* circuit, or perhaps a new type of circuit may be required to provide this linkage. Links from the radio ports to the central database may also be provided by private line data links. A more common arrangement, though, will use dial-up circuits to exchange information at specified periods.

INTERFACE TYPES

Depending upon the total voice channel requirements at a radio port site, the interface may be digital or two-wire analog.

CALL PROCESSING

At this time, it is unclear exactly how a CT2 system will function with the existing systems, (PSTN, cellular, etc.) because it is assumed CT2 systems in the United States will have some form of two-way capability. It is quite possible that the operation will be similar to that of existing cellular networks. In this case, the user is identified via information transmitted by the mobile unit. Routing for roaming users is accomplished through the use of a *Temporary Local Directory Number (TLDN).*

Mobile-To-Land Calls

To make a call using a CT2 system, the user first would have to determine that he/she is within one of the service islands of the designated CT2 provider. These islands could be identified by signs in public areas, or simply known to the user if the service is provided in the user's neighborhood. Within range of one of the radio ports, the subscriber would receive a signal that service is possible. The caller presses a button on the set, which transmits the ESN, MIN, and SID, to initiate a call. The radio port receives the signal and passes it to the radio controller, which verifies the authenticity of the user by comparing the data with information stored in the radio controller, or obtained elsewhere. Assuming the user is valid, the radio controller allocates a radio channel so the subscriber can dial the desired digits. Following receipt of these digits, the radio controller seizes an available dial line connection, and the call is processed through the PSTN.

If the user is a subscriber of another CT2 carrier, service is still possible. However, the radio controller would have to verify the user with information obtained from the other carrier. This information could be obtained, using *common channel signaling (CCS)* from another database or directly from the other CT2 carrier.

Land-To-Mobile Calls

In order for land-to-mobile service to be provided with a CT2 system, the location of the mobile subscriber would have to be known. One possible method would be for the mobile set to register automatically with a radio port, and that information sent back to the radio controller or central database. Calls destined for the mobile user would query the central database, or its equivalent. This query might require the use of a CCS link if the call involved more than one system. The database would let the originating system know the location of the user and provide a TLDN for routing purposes. The call would then be routed through the network to reach the radio port currently serving the subscriber.

PERSONAL COMMUNICATIONS NETWORK (PCN) SERVICE EVOLUTION

Similar to CT2, the *Personal Communications Network (PCN)* was first initiated in the United Kingdom. In 1989, the DTI also authorized three licenses to provide a new advanced personal communications service concept. These licensees were consortia known as Mercury PCN, Unitel, and British Aerospace (later called Microtel).

The exact technical standards that will be used to operate these systems are still under discussion. After considerable discussion, it appears that a standard based on the *Groupe Speciale Mobile (GSM)* protocol will be used. The main purpose of the additional licenses is to promote competition with the existing cellular systems.

A number of technical and market trials are currently underway in the United States to test the PCN concept. The participants have included Bell Regional Holding Companies, cellular operators, *Cable TV (CATV)* firms, and others.

PCN RADIO SPECTRUM REQUIREMENTS

Radio spectrum has not yet been allocated in the United States for PCN applications. However, PCN America petitioned the FCC in 1989 for a spectrum allocation in the 1700-2300 MHz band for PCN. It is possible that the FCC may decide to wait until the conclusion of the World Administrative Radio Conference in 1992 (WARC '92) before allocating any spectrum. This is because the WARC '92 may recommend a frequency band that can be used on an international basis.

In the United Kingdom, no specific frequency band has yet been reserved for the PCN providers. A recommendation has been made to utilize frequencies within the 1.71-1.88 GHz range, and this range will also be the U.K.'s recommendation at WARC '92.

Until the FCC makes a spectrum allocation, and rules on other aspects of PCN service, it is difficult to know how much spectrum each operator will be allowed to use. Assuming 100 MHz is allocated, it is believed that a PCN system could accommodate 10,000,000 users in a given area.

BASIC PCN ARCHITECTURE

In many ways, the attributes of a PCN are almost identical to those of a cellular system. Still to be determined is exactly how these PCN systems will be constructed. While it is possible that an entirely new network infrastructure will be developed, a more probable scenario is alliances with companies that already possess some infrastructure. These companies may be LECs, CATV firms, *interexchange carriers (ICs),* other utilities, or *competitive access providers (CAPs).*

The PCN concept is based on the use of low-powered transceivers located in very small areas called microcells. Compared to most cells in a cellular system, these microcells will cover a much smaller area. Cells in a cellular system generally cover areas that range in size from 1 mile to 50 miles radius or more. The microcells in a PCN system are expected to cover areas with a radius of about 200 to 2500 feet. Figure 8.2 illustrates the PCN concept.

Similar to the CT2 system, it is expected that services utilizing a PCN will employ handsets that transmit at about 10 milliwatts, or about 1/100th of a watt. It is possible that power levels as low as 1 milliwatt, or 1/1000th of a watt, will also be used. The base stations for the PCN system will generally be designed to transmit at 5 watts, or less, versus 100-500 watts ERP in existing cellular systems.

Handoff between cells will certainly occur in a PCN system in order to provide for frequency reuse. At this time, it is unclear whether these handoffs will accommodate high-speed vehicular traffic. It is possible that handoffs will be restricted to pedestrian, or slow-moving vehicular traffic. This will lower switching costs by reducing the processing power dedicated to administering handoffs.

As with existing cellular systems, the microcells will be linked to a central switch. However, it is expected that, unlike present cellular systems, the microcells will have some call processing capabilities. These will include processing calls within the same cell and perhaps performing handoff functions with adjacent cells. Having some intelligence in the microcells will reduce facility and switching costs for the PCN system.

Because of the lower power requirements, smaller antenna size, and perhaps limited handoff capability, many PCN providers believe they can offer a service that is considerably cheaper than current cellular service. Whether this cost savings can be achieved is debatable and very much dependent upon the cell sites. While it is anticipated that the PCN cell sites will be simple and inconspicuous, relatively huge numbers of sites

Personal Communications Network (PCN) Concept
FIGURE 8.2

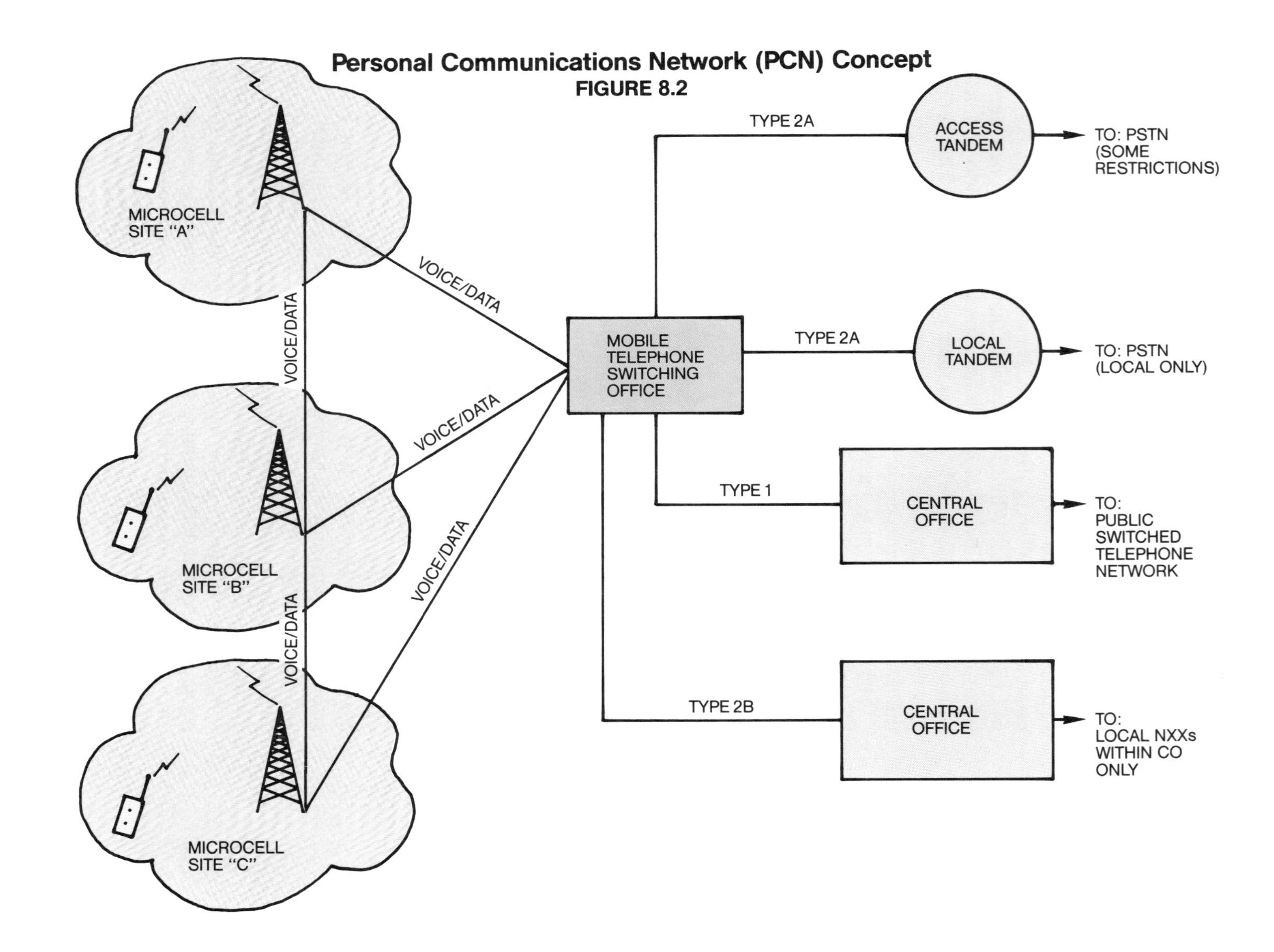

will be required in order to provide coverage for large geographical areas. Obtaining site locations can be difficult and expensive. In addition, it is important to remember that while existing cellular systems use much larger cell sizes and power levels, these systems have the capability to readily change both of those characteristics. Some cellular carriers are also proposing to offer a PCN-like service using their existing cellular frequencies.

PCN GRADE OF SERVICE

It is expected that PCNs will provide a grade of service that is comparable to existing landline or cellular service. Therefore, they should be engineered for a blocking probability of B.01 or B.02. This means that of every 100 attempted calls during the busiest period of the day, 98 or 99 will be completed.

PCN NUMBERING REQUIREMENTS

The exact numbering system to be used for PCNs has not been determined, although it is part of the FCC's inquiry into personal communication services. It is expected that PCNs will use numbers that conform to the NANP, whether it is the existing NANP or a new plan that is implemented in future years. The NANP number may be used as the *Mobile Identification Number (MIN),* like present cellular equipment.

Numbers used for PCNs may be from NXX codes shared with other services or may be dedicated to the PCN provider. Whether shared or dedicated NXX codes are used will depend on the numbering requirements of the provider and may depend on the interconnection type that is used. This is similar to the choices for cellular carriers described in Chapter 5.

INTERCONNECTION TYPES USED FOR PCN SERVICE

It is expected that PCNs will require trunk-side interconnection arrangements like those used by present cellular systems. These include Type 1, Type 2A, and Type 2B connections.

Private line links will also be needed to connect the microcells to each other, or back to the PCN equivalent of a *Mobile Telephone Switching Office (MTSO).* These links, if provided by the LEC, may be existing analog or digital (DS1 or DS3) offerings, or possibly fiber links. The PCN carrier may also obtain facilities from other sources, or construct its own facilities.

INTERFACE REQUIREMENTS

All of the switches used for PCN service are expected to be digital, so digital facilities will be used extensively. Regardless of the type of interconnection used, digital interfaces will be common at the *Point Of Termination (POT).*

Analog interfaces are available and are usually used when the number of trunks required is very small or when the PCN provider has a POT location in a remote area where digital facilities are not readily available. Because the transmission levels at the interface usually require some adjustment, channel terminating equipment needs to be installed by the LEC at the POT location.

It is also possible that an optical interface may be used in the future for some, or all, of these services.

Digital Synchronization

As with digital interconnection arrangements with other wireless carriers, an agreement must be reached between the PCN provider and the LEC regarding the primary source of timing.

GLARE RESOLUTION

PCN providers will be utilizing trunk-side connections, like existing cellular interconnection arrangements. This means an agreement will also have to be reached to determine who yields first in the event of a glare situation.

CALL PROCESSING

Standards for call processing with a PCN system have not yet been finalized. It is quite possible that a process similar to that of cellular systems will be used. This will require PCN systems to be able to know the location of mobile units and identify users accessing the PCN network. Therefore, information equivalent to the cellular MIN, ESN, and SID must be electronically encoded and made available to the PCN provider. It may also require the PCN provider to use a TLDN for users who have roamed to other areas or systems. Databases will be needed to store user location and other information. Information in these databases will be exchanged between other switches and systems using a CCS network with SS7 protocol.

Mobile-To-Land Calls

The mobile unit used for PCN service may indicate to the user whether service is available. It may also indicate to the system whether the user is "roaming" or not. After dialing the desired digits, the user would press the "send" button on the set. These digits would be transmitted to the base station on a control timeslot, along with the identifying data (MIN, ESN, and SID). Using the intelligence in the base station, voice channel availability is ensured, and the mobile unit is directed to a specified timeslot. The base station would then send the digits to the MTSO, which would seize a trunk to the PSTN or another network, according to the PCN provider's routing instructions. The called, or entered, digits would be outpulsed to direct the call to its desired location.

If the user was roaming, inquiries may be made to a database using the CCS link. A TLDN may also be assigned to the user's home system in order to route calls to the correct location.

Land-To-Mobile Calls

A PCN user would be called in the same manner as any user with a number that corresponds to the NANP. The caller would dial the 7- or 10-digit telephone number of the PCN subscriber, as required by local dialing practices. In order to complete the call, the location of the PCN subscriber would have to be known. This information could be verified by the CCS link before the call was forwarded through the network. If the PCN user was within the "home" system, the call would be directed to the appropriate connection (Type 1, 2A, or 2B), providing the interconnection with the PCN network.

If the user were "roaming," the call could be directed to the distant location using the TLDN provided by the home system upon inquiry with the CCS link. For example, suppose a PCN user was normally located in Philadelphia, but happened to be roaming in San Francisco. A call from Miami to the PCN user would not have to be routed to Philadelphia, only to be forwarded to San Francisco. Using information obtained with the CCS link, the call could be sent through the network directly from Miami to San Francisco, using the TLDN supplied by the San Francisco system.

MOBILE SATELLITE SERVICE

Satellites have been used for many years to provide point-to-point or point-to-multipoint voice, data, and video services. Nationwide paging, for example, is an existing form of personal communication services that utilizes satellite facilities. Another application is a marine satellite system *(Inmarsat),* which is used by thousands of ships and offshore drilling rigs. New applications have been proposed, which are labeled *mobile satellite service (MSS).* MSS may use different methods to provide service, but the intent is to enable users to have communication services available virtually anywhere in the world.

Domestic Mobile Satellite Service

In August 1991, the FCC tentatively decided that MSS in the United States would be provided using a consortium rather than using a comparative hearing process to select the licensees. Using a consortium allowed the FCC to expedite the entry process while simultaneously coping with the considerable costs of providing service via satellites.

The consortium is known as the *American Mobile Satellite Consortium (AMSC)* and currently comprises eight companies. The FCC has granted AMSC authority to construct and operate a domestic MSS system using the 1545-1559 MHz and 1646.5-1660.5 MHz bands. The first set of frequencies represents those used for *uplink* transmissions (earth to satellite) while the second set is used for *downlink* (satellite to earth) transmissions.

SATELLITE NETWORK ARCHITECTURE

Presently, the architecture of a satellite system is very similar to that of a two-hop microwave system. A transceiver, often called a dish because of its shape, is at one location and pointed towards the satellite in the sky. The satellite acts as a repeater because it receives the signal transmitted on the uplink from one transceiver location, changes the frequency of the signal, and retransmits the signal on the downlink to another transceiver site.

Satellites are an excellent means of providing coverage to large areas, particularly remote and sparsely populated areas. Their cost is relatively independent of the distances involved, they offer good digital performance, and they are fairly reliable.

Offsetting these virtues are several disadvantages of satellite service. Due to the physical limitations of the launch vehicles, the size and weight of satellites are limited. This limitation imposes some performance and capacity restrictions and greatly contributes to a very high initial cost. Satellites also have a limited lifetime, primarily due to the declining performance of the solar power cells and periodic positioning adjustments which use a fixed supply of on-board gas. For these reasons, satellites generally have an operational life of about seven to ten years.

Orbital Locations

Today's satellite systems use geostationary orbits in which the satellite appears to be stationary from any location on earth. This is accomplished by placing the satellite at an altitude of about 22,300 miles so that it revolves around the Earth over the equatorial plane at the same speed as the earth rotates. This simplifies tracking of the satellite by earth stations, but introduces a noticeable transmission delay and requires higher power levels because of the distances involved.

A new proposal would use *low earth orbit (LEO)* satellites to overcome the disadvantages of geostationary satellites. However, because of their low-altitude orbits, many satellites would be needed to provide the same coverage as a single geostationary satellite. The additional satellites are needed because a new satellite must come into view for a particular coverage area before the first satellite disappears over the horizon. While LEO satellites would permit the use of much smaller user terminals, the complexity of the system is increased due to the need to track each satellite and provide handoffs between satellites to avoid interrupting calls.

Motorola has proposed an LEO satellite system called Iridium to provide worldwide coverage. The system would require more than 70 satellites and cost more than $2 billion to construct. But it would provide service at any place on the globe to as many as 10 million subscribers. At this time, the proposed Iridium system is not to be part of the domestic MSS which the FCC authorized in 1991.

No spectrum allocations have yet been made by the FCC for LEO satellite operation. Unlike geostationary systems, frequencies used for LEO operation must be coordinated on an international basis because they pass over all parts of the world. Because geostationary satellites do not move with respect to the earth, coordination is not needed for countries located on the other side of the globe.

MOBILE SATELLITE SERVICE NUMBERING REQUIREMENTS

Numbering requirements for MSS are currently unclear. However, for these services to provide true mobility for customers, these systems must possess the ability to utilize numbers that conform to the *North American Numbering Plan (NANP).* Whether these numbers are from shared or dedicated NXX codes will depend on the demand for this service.

INTERCONNECTION TYPES USED FOR MSS

Some existing satellite-based services interconnect with the PSTN. These connections are typically dial line connections. It is possible that future satellite services will require trunk-side connections, like the Type 1, Type 2A, or Type 2B connections widely used by existing cellular carriers.

INTERFACE TYPES

Existing interconnection arrangements are primarily analog due to the small number of connections required. Future arrangements may use analog or digital interfaces, depending on the total voice channel requirements of the MSS.

CALL PROCESSING

Standards for call processing using an MSS or an LEO satellite system such as Motorola's Iridium proposal have not been finalized. It is likely that these systems will process calls in a manner similar to existing cellular systems, particularly if the system uses a handoff capability between satellites. Also, in order for these systems to exchange location and user information with other wireless systems, database information will have to be exchanged using a CCS network with SS7 protocols, or an equivalent arrangement.

9 REGULATORY PERSPECTIVE

All wireless services, or devices, utilizing radio frequencies are subject to some form of regulatory constraints. These constraints may be imposed by the federal government, via the *Federal Communications Commission (FCC),* by state agencies, or both. State regulations normally emanate from agencies that control public communication services. The exact names of these agencies vary widely, but they are often called *Public Service Commissions (PSCs).* Decisions by either type of agency can have a significant effect on the provision of wireless services.

FCC STRUCTURE

For the purposes of wireless services, as depicted in Table 9.1, the FCC allocates radio spectrum, determines whether a frequency band should be used for private or common carrier purposes, issues radio licenses, establishes certain interconnection policies, and approves rates for interstate services.

Table 9.1
Federal and State Regulatory Jurisdiction

FUNCTION	FCC	STATE
Allocate Radio Spectrum	YES	NO
Issue Radio Licenses	YES	NO
Establish Interconnection Policies	YES	NO
Private or Common Carrier Status	YES	NO
Interstate Interconnection Rates	YES	NO
Intrastate Interconnection Rates	NO	YES
Intrastate User Eligibility	NO	YES
Tariff or Contract Used for Interconnection	NO	YES

Initially, allocation of radio frequencies was handled by the Federal Radio Commission, which was established in 1927. As a result of the 1934 Communications Act, this function and others were assumed by the FCC. Decisions issued by the FCC are rendered by vote of the commissioners or by specific Bureaus within the FCC. There are four Commissioners and a Chairman, all appointed by the President and confirmed by the Senate. Within the FCC are four separate Bureaus, eight Offices, and a Review Board. These units constitute the permanent staff of the Commission. Each Bureau administers specific portions of the FCC rules, and is headed by a designated Bureau Chief. Various Divisions within each Bureau are assigned specific functions. For wireless telecommunication services, the most pertinent Bureau is the Common Carrier Bureau. A principal Division in the Common Carrier Bureau is the Mobile Services Division.

FCC DECISIONS

Decisions by the FCC result from petitions or complaints filed with the FCC, or by rulemaking proceedings initiated by the FCC. These decisions can be rendered by the full Commission, a Bureau, or a Division. The decisions have titles, such as *Order, Report and Order (R&O),* or *Memorandum Opinion and Order (MO&O).* An Order or a Report and Order usually results from a petition or a complaint filed with the Commission. A Memorandum Opinion and Order is normally issued in response to a rulemaking proceeding. Based on the authority granted to the FCC by Congress, they all have the same force and effect of law, although the full Commission can overturn a Bureau or Division decision.

Sometimes the FCC will seek comments on a particular topic of interest by initiating a proceeding called a *Notice Of Inquiry (NOI).* After gathering data, the FCC may decide to propose rules through another proceeding called a *Notice Of Proposed Rulemaking (NPRM).* Eventually, a decision in the form of an Order, Report and Order, or Memorandum Opinion and Order may be issued to complete the process.

FCC RADIO SPECTRUM ALLOCATION AND LICENSING

A major duty of the FCC is to allocate radio spectrum and establish rules to prevent radio interference between users. Therefore, any radio device in the United States must meet certain FCC criteria. In some cases, such as consumer products, the rules only apply to the manufacturer. In the case of wireless services, the amount of regulation is usually considerably greater. The applicable rules of the FCC are enumerated in exacting detail in Title 47 of the *Code of Federal Regulations (CFR).*

Portions of the radio spectrum are allocated for many purposes. Some of the frequencies, like cellular, are reserved for the exclusive use of the authorized licensee. Other services, like certain dispatch services, share frequencies.

The number of entities utilizing a particular frequency band to provide service may be limited by FCC statute, or simply by frequency availability. Cellular service has a defined number of licensees, two for a given geographical area. For other services, like paging that is provided on a common carrier basis, the frequencies are assigned to eligible carriers on a first-come, first-served basis. In this instance, the number of carriers in a given geographical area is only limited by the available spectrum.

COMMON CARRIERS AND PRIVATE CARRIERS

Some radiotelecommunication services may be provided by a *common carrier* or a *private carrier.* The specific rules for each classification are contained in Part 22 and Part 90 of the FCC rules, respectively. For this reason, these carriers are often referred to as Part 22 or Part 90 carriers. Some services, like cellular and air-ground, may be provided only by common carriers. Other services, such as two-way mobile and paging, can be provided by either common or private carriers. The differences between common and private carriers are regulatory in nature, and not nearly as clear as they once were.

Common Carriers

Common carriers provide public services, so they normally cannot deny service to people who are willing to pay for the services that are offered by the carrier. They are also subject to state regulation so their ability to change their rates may be limited. The FCC licensing process is usually longer for a common carrier than for a private carrier because much more information is required.

Offsetting these disadvantages for the common carrier are several positive things. First, the common carrier has exclusive use of the frequencies assigned for its use. Second, it is allowed to make a profit on calls that use the carrier's interconnection facilities with the *Public Switched Telephone Network (PSTN).* Finally, the end user does not have to apply for an FCC license in order to operate the radio equipment, like a cellular phone, which is used with the common carrier's service.

Private Carriers

Private carriers serve specific categories of end users. Examples of these are Business, Petroleum, Public Safety, Forest Products, and Government. In effect, virtually any individual qualifies as an end user under some category in Part 90 of the FCC rules. Private carriers enjoy several advantages not available to a common carrier. They can be selective with respect to their customer base because they do not have to serve everyone within a particular category. An additional advantage is that a private carrier's rates are not subject to state regulation. The FCC licensing procedure is also much simpler and quicker for private carrier operation. Finally, they have more flexibility in financial arrangements because they do not have the foreign ownership constraints imposed on common carriers.

There are also disadvantages for private carriers. The biggest drawback is the fact that private carriers must share frequencies with other private carriers. Furthermore, a private carrier's equipment must not cause interference with the other private carriers. Also, while a private carrier may interconnect its system with the PSTN, it cannot make any profit on the traffic that uses this interconnection arrangement. Moreover, the FCC has not mandated any particular form of interconnection, such as Type 2 arrangements, as it has for common carriers. For many private services, like dispatch and two-way mobile applications, the end user must obtain an FCC license for the equipment he/she operates.

FCC INTERCONNECTION POLICIES

Over the years, the FCC has issued a number of decisions that have had a significant effect on interconnection arrangements. The intent of each of these decisions has been to promote competition for wireless services. In chronological order, the major decisions are summarized in the following paragraphs.

1949—Creating The Radio Common Carriers (RCCs)

On September 30, 1946, the FCC authorized a pair of frequencies in the 150 MHz band for the exclusive use of a *non-wireline* entity. Non-wireline meant it was not affiliated with any entity providing telephone service as a *Local Exchange Carrier (LEC).* These new entrants were initially labeled *Miscellaneous Common Carriers (MCCs),* but later became known as *Radio Common Carriers (RCCs).* However, this initial frequency allocation was for experimental purposes only.

In 1947, the FCC began another proceeding in Docket 9046. This culminated in a decision on May 11, 1949, which authorized the new RCCs to provide mobile services on a competitive basis with the LECs. To enhance this effort, the FCC authorized specific frequencies for the exclusive use of the RCCs. However, the FCC did not address the interconnection requirements. While the RCCs did obtain interconnection arrangements with some LECs, they were often denied the same arrangements used by their LEC competitors.

1968—The "Guardband" Decision On Competitive Equality

This decision is referred to as the *Guardband* decision because it concerned the allocation of certain frequencies, formerly used for guardband purposes, to provide additional channels for paging services. Guardband frequencies are used as buffers between working channels to reduce adjacent channel interference.

As a result of its decision in Docket 16778, issued on May 8, 1968, the FCC ordered the LECs to provide dial access arrangements to the RCCs, upon request, in any community where such arrangements were being used by the LECs. Furthermore, it required the LECs to charge the RCCs for facilities using the same cost computations used for the LEC's own competitive services.

1981—Establishing Cellular Service

Several important principles were contained in a Report and Order issued on April 9, 1981, in Docket 79-318. To further promote competition, the FCC created geographical areas based on the *Metropolitan Statistical Areas (MSAs)* defined by the U.S. Census Bureau. There are 305 MSAs in the United States plus a special MSA, Number 306, created to serve offshore areas in the Gulf of Mexico. Within each MSA, two licensees would be authorized to provide cellular service. One would be affiliated with a LEC and called the *wireline* carrier. The other licensee would be unaffiliated with a LEC and labeled the *non-wireline* carrier. Since that time, due to many mergers and acquisitions, many companies are the wireline carrier in one MSA and the non-wireline carrier in another. In any event, a carrier cannot serve as both the wireline and non-wireline carrier in the same MSA.

The FCC also recognized that cellular carriers were also common carriers, and not end users. Cellular systems were to be fully interconnected with the PSTN through arrangements most favorable to the end users. While the FCC noted that the *Mobile Telephone Switching Office (MTSO)* could perform end office functions, it did not mandate a tandem-type interconnection arrangement. But, any interconnection arrangement for the non-wireline carrier was to be on terms that were no less favorable than those offered by the LEC to the wireline carrier.

1986—Cellular Policy Statement

Because the FCC had received a number of complaints about difficulties negotiating interconnection arrangements with the LECs, a policy statement was issued on March 5, 1986. This Memorandum Opinion and Order (FCC 86-85), reiterated some of the principles outlined in the 1981 order, but also formalized some other positions.

The FCC reiterated that the LECs, at a minimum, must provide the non-wireline carrier with interconnection arrangements that were on terms no less favorable than those provided to wireline carriers. It also recognized that the interconnection arrangements may differ and ordered the LECs to provide tandem interconnections, labeled Type 2, if requested by the cellular carrier.

Once again, cellular carriers were determined to be common carriers. The FCC expanded its previous rulings and stated that the cellular carriers provided local exchange-like service. Therefore, they were *co-carriers,* not end users. However, the compensation arrangements used for co-carriers was declared to be largely a matter for state jurisdiction.

In some cases, obtaining telephone numbers from the LECs had been very difficult for the cellular carriers. The FCC noted that the LECs did not "own" the numbers and that the cellular carriers were entitled to numbers on the same basis as another LEC.

1986—Establishing The Rural Service Areas (RSAs)

With the 306 MSAs, cellular service would be available in the major metropolitan areas of the United States. However, a large portion of the geography, and a number of important smaller cities, were not part of the MSAs. Using a plan based on county boundaries, the FCC created 428 *Rural Service Areas (RSAs).* This decision, issued on June 26, 1986, in Docket 85-388, contained a number of rules that were designed to prevent speculation for the licenses. For interconnection purposes, a major difference between the MSAs and the RSAs is that the license in an RSA can be shared. This results in multiple operators within an RSA although only two licenses, wireline and non-wireline, are actually issued.

1987—Declaratory Ruling

About a year after issuance of the policy statement, the FCC expanded its interconnection policy with a Declaratory Ruling issued in Docket 87-163 on April 30, 1987. With this ruling, the Commission extended the provisions of its 1986 Policy Statement to all Part 22 common carriers, not just the cellular carriers. As a result, Part 22 carriers providing paging, two-way mobile, air-ground, and offshore service were now entitled to the same interconnection arrangements as cellular carriers.

In addition, the FCC ruled on other aspects of interconnection. It declared that tandem (Type 2) connections should generally be cheaper than end office (Type 1) connections. The FCC also outlined the *mutual compensation* principle, which said wireless carriers should be compensated for the switching costs they incur when traffic is terminated on their networks. Mutual compensation was applicable only for Type 2 connections, and only for interstate traffic.

1990—Personal Communications Service (PCS) Inquiry

In June, 1990, the FCC began seeking comments on Personal Communications Service (PCS) in Docket 90-314. Of particular interest is a definition of PCS, system standards, spectrum allocation needs, provider eligibility requirements, and determining whether a new numbering plan is needed for PCS.

In October, 1991, a Policy Statement was issued which indicated that further information would be sought via an *En Banc* hearing with all of the Commissioners. It proposed focusing on the 1850-2200 MHz frequency band, but did not exclude other bands. The FCC stated that the initial spectrum allocations should occur in 1992. No determination was made as to whether the PCS providers would be classified as common carriers, private carriers, or both. It is difficult to predict when the FCC will render a final decision to conclude this docket.

STATE REGULATION

Because the FCC has issued many decisions to promote competition, it would appear that the role of the state regulators is rather minimal. Indeed, the FCC does play a major role, but equally important is the role of the state regulatory commissions. This is because the states have the right to regulate the subscriber rates charged by wireless carriers (at least of the common carrier variety). This is especially important because the bulk of the wireless carrier's traffic is normally intrastate, not interstate. The amount of state regulation that is imposed varies from practically zero to considerable, depending on the state.

State Regulatory Organization

The composition of a state regulatory commission varies by state. Many have five Commissioners, just like the FCC. In previous years, these Commissioners were elected officials. More recently, most states are appointing Commissioners in a process that also resembles the FCC's, but involves only the state government.

Each state commission also has a staff that analyzes issues and makes recommendations to the Commissioners. There is considerable variation among the states in the size of a state staff, and their aggressiveness in pursuing issues.

As shown in Table 9.1, the states can neither allocate radio spectrum nor issue radio licenses to operators. So, the states do not select which common carriers will be allowed to provide wireless services within their territory. But, the states do have the right to establish the business conditions, or regulatory climate, for these common carriers.

The states must adhere to the interconnection policies established by the FCC, but only as long as these policies do not pertain to intrastate rates. For example, the states cannot deny the provision of Type 2 connections to common carriers because these connections can be used for both intrastate and interstate traffic, and the two are inseparable. Yet, while the FCC has said that Type 2 connections should generally be cheaper than Type 1 connections, this is essentially a rate issue. Hence, a number of states, like Alabama and South Carolina, have decided that charges for Type 1 and Type 2 should be identical for intrastate purposes.

Another example is the mutual compensation principle promulgated by the FCC. While this is applicable for interstate traffic, states such as California and Wyoming have determined that it is inappropriate for intrastate traffic.

States may elect to expand the interconnection policies of the FCC. For instance, the 1987 Declaratory Ruling said the FCC's interconnection policy applied to all Part 22 carriers, but the FCC has not extended this policy to include the private (Part 90) carriers. Although states cannot regulate the private carriers, they can exercise jurisdiction over the interconnection arrangements offered by the LECs. At least one state, Florida, has determined that the same interconnection arrangements shall be offered by the LECs to both common and private carriers.

Whether the interconnection arrangements for intrastate service result in a tariff filed by the LEC, or in a contract between the LEC and the wireless carrier, is also a state matter. Regardless of which is used, both parties must attempt "good faith" negotiations, as ordered by the FCC. However, even though the FCC has said contractual arrangements are preferable, the states have the right to require a tariff filing. Even if contracts are used, some states require regulatory approval before these contracts become effective. Other states simply require the contracts to be filed with the commission and take no other action.

The degree of regulation which states impose on the wireless carriers themselves is also considerable. Some states require the wireless carriers (wireless common carriers only) to file tariffs that detail the charges for the services they are offering to the public. Some states insist that tariff changes be filed before price changes can become effective while others allow changes within certain parameters without any tariff filings. Some states only require the wireless carrier to obtain a *Certificate of Public Convenience (CPC)* and allow the rates to be established by the market.

Presently, about half of the states regulate common carriers which provide two-way mobile or paging services. Somewhat less than half of the states regulate cellular carriers.

GLOSSARY

2-way trunk: A trunk that can be seized at either end.

4-wire circuit: A circuit using separate transmission paths for each direction of transmission.

Analog signal: A signal that changes in a constant fashion, such as voice.

Automatic Number Identification (ANI): The number of the calling station that is automatically forwarded by a switching system.

Billing Telephone Number (BTN): A telephone number used for billing purposes rather than identifying the actual calling station.

Clock: A source of timing reference for digital equipment.

Common Channel Signaling (CCS): A signaling method utilizing a separate path from the voice path which carries signaling messages pertaining to a number of circuits.

Decibel: A unit for expressing the ratio of two amounts of electric or acoustic signal power equal to 10 times the logarithm of this ratio.

Dial Pulse (DP): Changes in current flow, such as a rotary dial telephone, that provide address (or dialed digits) information.

Dial tone: An audible signal from a switching system that indicates to a customer that the equipment is ready to receive dial signals.

Digital rate: The number of bits transmitted per unit of time.

Digital Reference Signal (DRS): A sequence of digital bits that represents a 1004 Hz to 1016 Hz signal with a power level equal to one milliwatt.

Digital signal: A signal that has a limited number of discrete states.

Digital signal level: One of several transmission rates in the time-division multiplex hierarchy.

Digital transmission: A mode of transmission in which all information to be transmitted as a stream of pulses.

DS1: A digital signal at the rate of 1.544 Mbps that is the equivalent of 24 analog voice channels.

DS3: A digital signal at 45 Mbps that is the equivalent of 28 DS1 signals, or 672 analog voice channels.

Dual-Tone Multifrequency (DTMF) signaling: A method for transmitting address pulses, and other signaling information, using a set of dual-tones to represent individual characters or numbers. It is primarily used for signaling by station equipment, like telephone sets.

Effective Radiated Power (ERP): An expression of the signal power which is actually radiated. It is the power of the transmitter, minus any transmitter cabling loss, plus any antenna gain.

Equal Access: A feature of some switching systems that provides the subscriber the ability to automatically select which Interexchange Carrier (IC) will carry their interLATA traffic without dialing any additional codes.

Full-duplex transmission: A communication circuit that can simultaneously transmit and receive information.

Half-duplex transmission: A communication circuit that can transmit or receive information, but not simultaneously.

High-usage trunk group: A trunk group that is designed to carry the majority of traffic between two switching systems, but has the capability to overflow excess traffic to a tandem route.

Inductor: An electrical device that can effectively shorten the required length of an antenna.

Interexchange Carrier (IC): A telecommunications common carrier authorized by appropriate regulatory agencies to provide service on an intraLATA or interLATA basis.

Interface: The point of interconnection between a wireless carrier's and a local exchange carrier's communication facilities.

InterLATA: Telecommunication services or functions that originate in one LATA and terminate in another LATA.

IntraLATA: Telecommunication services or functions that originate and terminate within the same LATA.

Line: A central office connection that usually connects the switching system to customer equipment.

Local Access Transport Area (LATA): A geographical area adopted at the Bell System divestiture to identify "exchange areas" to be served by the local exchange companies.

Local Exchange Carrier (LEC): A company which provides intraLATA telecommunications within a franchised territory.

Loop: A channel between a customer terminal and a central office.

Metropolitan Statistical Area (MSA): Sometimes known as Standard Metropolitan Statistical Areas (SMSAs), MSAs are areas based on counties as defined by the U.S. Census Bureau that are cities of 50,000 or more population and the surrounding counties.

Multifrequency (MF) pulsing: A method for transmitting address pulses, and other signaling information, using a set of dual-tones to represent individual characters or numbers. It differs from DTMF in that it uses different frequencies and is used for signaling between switching systems.

Multiplexing: The process or equipment for combining a number of individual channels into a common spectrum or into a common bit stream for transmission.

On-hook: An electrical indication that the equipment is in an idle state.

Off-hook: An electrical indication that the equipment is in a busy or a request for service state.

Point of Termination (POT): The physical location marking the point at which the local exchange carrier's service ends.

Presubscription: The process used by a customer, where equal access service is available, to select their designated interexchange carrier.

Public Safety Answering Point (PSAP): The public facility designated to receive and respond to emergency calls.

Read Only Memory (ROM): An electrical circuit that can store encoded information that is not expected to change.

Signaling System 7 (SS7): An internationally standardized protocol used for common channel signaling purposes.

Toll traffic: Traffic destined for areas beyond the normal local calling area of a local exchange carrier.

Transmission Level Point (TLP): A point in a transmission system used to determine the power of a test signal, in decibels, with respect to the power of a test signal at a reference point. For example, the power of a test signal at the +7 TLP should be 7dB greater than the power of a test signal at the 0 TLP.

Trunk: A communication path connecting two switching systems.

Wireless carrier (WC): A generic term used to describe entities providing wireless service. Such entities include, but are not limited to, cellular carriers, radio common carriers, and private carriers.

INDEX

ABOUT THE AUTHOR

Harry E. Young is on loan from BellSouth to Bellcore where he serves as a Director-Wireless Interconnection for Bellcore. In this capacity, he advises Local Exchange Carriers on wireless services interconnection matters, including technical requirements, operational considerations, and business implications of interconnection arrangements. Harry also has responsibility for a major technical reference for wireless interconnection, TR-NPL-000145, and serves as moderator of the Wireless Interconnection Forum (WIF). He has conducted a number of public interconnection seminars and has written several magazine articles on wireless interconnection. Harry also wrote Section 16 on Mobile Interconnection in Bellcore's *"BOC Notes On The LEC Network—1990."* He has 28 years of experience in the telecommunications field and has been involved with interconnection matters for over 20 years. Harry graduated from two fine academic institutions, but they represent the opposite ends of the college football continuum. He holds a B.B.A. degree from the University of Miami and a M.S. degree from Columbia University.